Federal Energy Administration Regulation

Federal Energy Administration Regulation

Report of the Presidential Task Force

Edited by Paul W. MacAvoy

Ford Administration Papers on
Regulatory Reform

American Enterprise Institute for Public Policy Research
Washington, D.C.

Paul W. MacAvoy is professor of economics at Yale University and an adjunct scholar at the American Enterprise Institute.

Library of Congress Cataloging in Publication Data

United States. Task Force on Reform of Federal
    Energy Administration.
    Federal Energy Administration regulation.

    (AEI studies ; 150)
    Includes bibliographical references.
    1. Petroleum industry and trade—United States.
2. Industry and state—United States.   3. United
States. Federal Energy Administration.   I. MacAvoy,
Paul W.   II. Title.   III. Series: American Enterprise
Institute for Public Policy Research. AEI studies ; 150.
HD9566.U56   1977   338.2'7'2820973   77-6697
ISBN 0-8447-3248-6

AEI Studies 150

© 1977 by American Enterprise Institute for Public Policy Research, Washington, D.C. Permission to quote from or to reproduce materials in this publication is granted when due acknowledgment is made.

Printed in the United States of America

CONTENTS

FOREWORD

Early in 1975, I called for the initiation of a major effort aimed at regulatory reform. Members of my administration, and the Congress, were asked to formulate and accelerate programs to remove anti-competitive restrictions in price and entry regulation, to reduce the paper work and procedural burdens in the regulatory process, and to revise procedures in health, safety and other social regulations to bring the costs of these controls in line with their social benefits.

My requests set in motion agency and department initiatives, and a number of studies, reorganization proposals, and legislative proposals were forthcoming last year. A number of these resulted in productive changes in transportation, retail trade, and safety regulations. Nevertheless, much remained to be done, in part because of the time required to complete the analysis and evaluation of ongoing regulations.

This volume provides one set of the analytical studies on regulatory reform that were still in process at the end of 1976. Necessarily, these studies would have undergone detailed evaluation in the agencies and the White House before becoming part of any final reform program. They do not necessarily represent my policy views at this time, but they do contribute to the analyses that must precede policy making. I look forward to the discussion that these papers will surely stimulate.

Gerald R. Ford

GERALD R. FORD

PREFACE

In February of 1976 President Gerald R. Ford asked his regulatory reform group (the Domestic Council Review Group on Regulatory Reform) to establish task forces to improve the operations of the regulatory agencies under his jurisdiction. This request was made when the Ford administration's proposals for legislative reform of transportation, energy, and financial regulations were before Congress, and to make a good case for such reform in the independent agencies called for improving operations in the executive office agencies. The effort was to be directed towards removing complicated, redundant, and costly procedures where there were simpler and more direct ways of achieving the same goals in regulation. After extensive surveys of the agencies, offices, bureaus, and commissions in the executive office that carry on regulatory activities, the reform group decided that a priority effort required a task force in the Federal Energy Administration (FEA) to improve on the complicated allocation and price controls that applied to refined petroleum products.

On the basis of an investigation and analysis stretching over the last half of 1976, the Task Force on Reform of Federal Energy Administration Regulation concluded that FEA regulation imposed significant costs on markets for refined petroleum products. The petroleum industry paid reporting and administrative compliance costs approaching $500 million annually, and firms were subject to extensive governmental interference with the conduct of normal business activities. In addition, taxpayers paid $47 million or more each year to maintain FEA's regulatory program. For all this outlay of time and money, consumers did not receive the benefits Congress intended to be obtained in refining from controls on crude prices, because FEA regulations (1) discouraged the construction of new refinery capacity necessary to meet future demand for refined petroleum products, (2) encouraged inefficiencies in the distribution of petroleum products, (3) created barriers to the importation of cheaper foreign products, and (4) created incentives toward utilization of domestic refining capacity at levels where incremental costs of production and prices to consumers were substantially increased.

These were the main results from FEA controls when and where complied with. But the task force found that compliance could not be counted on because the FEA compliance program was ineffective. To enforce the regulations required information about prices during the base period of May 1973 and about add-ons since then, but FEA had not yet determined the May 1973 base prices. Moreover, the allocation regulations could not be effectively enforced because the base period relationships no longer reflected current market supplier/purchaser relationships and base period volumes had been so overstated that allocation entitlements were 30 percent above current usage for some fuels. What was ineffective under current market-clearing conditions might be quite effective in an embargo, however. In fact, the task force found that the regulations would not work in a shortage because of the misallocations from enforcing outdated and irrelevant controls and because sanctions were insufficient to deter violations.

Many exceptions had been granted to the May 1973 patterns of controls, as well. The exception process was found to have evolved into a mechanism that not only provided case-by-case relief but also developed rules of general application. Because of this rulemaking the process created vested interests in continued regulation and changed the nature of regulatory policy without a conscious agency decision to do so.

On the whole, the task force found that the cost of current FEA regulations had come to outweigh their benefits. Moreover, more important controls—that is, the product pricing and allocation regulations, the crude oil buy/sell program, and the supplier/purchaser freeze—were unnecessary in present supply conditions. In the long run, continuation of these controls would result in substantive inefficiencies in the industry. On the basis of these findings, the task force recommended that during a normal supply period the product price and allocation regulations for refiners and resellers should be eliminated, as should the buy/sell program and the supplier/purchaser freeze for crude oil. At the same time, the entitlements program should be broadened to include imported products as well as crude oil, and the small refiner bias should be eliminated.

But the task force also recommended that during periods in which there were interruptions of supply or a shortage, FEA should adopt a regulatory "trigger mechanism" to commence and terminate a set of stand-by controls. The stand-by controls themselves should be amended to make them workable and enforceable in a shortage as follows: (1) update base periods; (2) require a single maximum selling price within a Petroleum Administration for Defense (PAD) district for each product of each refiner, based on base period prices plus full passthrough of both product and nonproduct costs for refiners; (3) freeze reseller margins but allow full passthrough of increased product costs; (4) eliminate class-of-purchaser price rules; (5) retain the crude oil buy/sell program without the small refiner advantage; (6) eliminate the supplier/purchaser freeze on crude oil; (7) retain the supplier/purchaser freeze on products; (8) retain the entitlements program for crude oil without the small refiner bias and establish a product entitlements program without a small refiner bias.

These proposals are central to dealing with the problems of costly regulations in energy, and they have to be addressed as part of the process of formulating an effective national energy policy. The task force report providing the documentation and analysis of these findings and recommendations is reproduced here with only marginal deletions as an important contribution to education for that critical process of policy formation.

PAUL W. MACAVOY
Yale University
April 1977

1

FEA REGULATION OF THE PETROLEUM INDUSTRY

In this chapter of the report, the task force will attempt to lay a foundation for its evaluation of the petroleum industry regulations administered by the Federal Energy Administration (FEA). The first part of the chapter will briefly describe the nature of the petroleum industry and the status of competition in it. Then it will note the impact upon the industry of government regulation and will outline the genesis of the current regulatory structure. The second part will describe various regulatory "programs" which comprise the regulatory structure which FEA currently administers.

The Nature of Competition in the Petroleum Industry and the Impact of Government Regulation

The Nature of Competition in the Petroleum Industry. The petroleum industry is a complex network of more than 300,000 firms responsible for extracting vast quantities of crude oil from underground reservoirs, refining it into usable products, and distributing it to millions of users. The firms in the petroleum industry perform five basic functions: (1) exploration and production, (2) gathering and crude purchasing, (3) transportation, (4) refining, and (5) marketing.

Firms operating in the petroleum industry are of greatly varying sizes. At one end are the fifteen to twenty-five major integrated petroleum companies, such as Exxon Company USA and Texaco, which engage in all functions performed by the industry. At the other end are the many thousands of independent companies whose activities are confined to just one activity within the industry.

In order to analyze the effect of FEA regulations on the petroleum industry and the economy, it is necessary to make certain judg-

1

ments about the nature of competition among these firms, absent government regulation. The task force has made the assumption that at present the domestic petroleum industry is workably competitive under conditions of adequate crude oil and product supplies. We recognize that there is controversy about this subject.[1] We note, however, that Congress, in passing the Emergency Petroleum Allocation Act (EPAA) of 1973, and recently in concurring in the decontrol of residual fuel and several refined products, has adopted the assumption that the petroleum industry is workably competitive. The task force also believes that this assumption is the appropriate one to make on the basis of current economic evidence. The current public record shows that concentration ratios, profit levels, and the relative lack of barriers to entry are in accord with those normally found in workably competitive industries. We also find that in times of adequate supplies, as currently exist, nonintegrated firms in the petroleum industry are not, as a class, at a competitive disadvantage with respect to integrated buyers, except insofar as integration may provide genuine economic efficiencies.

The task force notes that the use of pipelines to gather and transport crude oil and to transport refined petroleum products does give rise to monopoly power on the part of those controlling the pipelines. This monopoly power results from the fact that efficiencies in pipeline transportation increase continuously with size, making the construction of competing pipelines serving the same routes economically inefficient. Fortunately, most intrastate petroleum pipelines and all interstate petroleum pipelines are regulated as common carriers, with interstate pipelines being under the jurisdiction of the Interstate Commerce Commission. As such, they have an obligation to ship petroleum on nondiscriminatory terms for all who wish to have their petroleum transported.

Domestic production of crude oil is no longer sufficient to meet domestic needs. Consequently, the marginal barrel of crude oil needed to supply increasing domestic demand must now come from foreign, rather than domestic, sources. At the present time, there are no limitations on the total volume of petroleum which may be imported; under the Mandatory Oil Import Program (MOIP), unlimited volumes of crude oil may be imported. Even without price controls, the domestic petroleum producers could not act as a cartel to set the price of petroleum above the world price, and the price of petroleum would be equivalent to the landed price of a barrel of imported crude.

[1] See, for example, S. Rep. No. 94-1005, 94th Congress, 2nd session (1976), pp. 4-5 (Petroleum Industry Competition Act).

Any attempt by domestic producers to charge more than the world price for crude would simply result in increased imports of crude at the world price. Consequently, domestic producers would, in the absence of regulation, take the world price as given, and they would offer for sale those quantities of crude whose production would be economically justified at the world price.

Of course, the world price is today set by the Organization of Petroleum Exporting Countries (OPEC) through a cartel agreement among producing governments. This situation does not, however, change the fact that without regulation domestic producers would charge whatever price was set for world petroleum, no more, and no less.

Effect of Prior Government Regulation. In analyzing the regulations imposed on the petroleum industry during the last five years, it is necessary to remember that the petroleum industry—or, more specifically, the crude production segment of the industry—has been subject to pervasive government regulation for the past forty years. Unlike the current price regulations which are designed to hold prices down and transfer income from the petroleum industry to consumers, previous regulatory programs had just the opposite effect: to stabilize and enhance domestic petroleum prices and transfer funds from consumers to petroleum producers.

In the 1930s, an overproduction of petroleum drove down the price of crude oil in many Texas and Oklahoma fields to a level of about ten cents per barrel. The governors of those two states declared martial law in several fields and shut down production. In 1935, major producing states signed the Interstate Oil Compact to Conserve Oil and Gas.[2] In implementing this compact, the major producing states instituted programs which had the effect of both enhancing the long-run production of petroleum and stabilizing prices. Most importantly, the major producing states developed a system of "prorationing" under which producers were permitted to produce only a specified percentage of the maximum efficient rate (MER), an amount established by state conservation agencies as the rate of production which would maximize the long-run output of a petroleum reservoir. Under the prorationing programs, state agencies, such as the Texas Railroad Commission, set the rate of "allowables" at a level designed to meet expected demand. Of course, the fixing of a maximum output for all producers had the effect of stabilizing prices, for,

---

[2] 49 Stat. 939 (1935).

inherently, a given level of production establishes a price which will clear the market at expected levels of demand.

The institution of prorationing also had the effect of eliminating competition among producers of crude oil, since they could generally expect to sell all that they could produce at the level set by the state and they did not have to lower their price to attract purchasers. Conversely, purchasers of crude were aware that they could not receive a lower price by attempting to shop among producers, and so they had no incentive to bid for crude on the basis of price.

It should be noted that, as long as the level of permissible production was set below 100 percent, additional domestic supply could be obtained merely by convincing a state regulatory agency to increase the level of production in that state. As a result of this circumstance, independent refiners were usually assured of an adequate supply of crude oil, since most state regulatory agencies would give favorable consideration to a complaint by a refiner that it was unable to obtain crude at the current price. The state agency would increase the level of allowables, and the increased production would provide the petroleum inputs for the refiner.

Congress lent the weight of the federal government to the enforcement of state prorationing schemes with the passage of the Connally Hot Oil Act.[3] The statute made it unlawful to ship in interstate commerce petroleum produced in excess of that permitted by a state conservation agency.

In the mid-1950s, the increasing availability of low-cost foreign crude threatened to undermine the domestic prorationing system and undercut the relatively stable prices that had resulted from the prorationing system. Officials both in Congress and in the executive branch were concerned that increasing imports of foreign petroleum would undercut the viability of domestic petroleum exploration and production efforts to the extent that supplies of petroleum adequate to meet the national defense requirements of the United States might no longer be available. In 1959, therefore, President Dwight D. Eisenhower, by proclamation, established the Mandatory Oil Import Program, which set volumetric limits on the amount of crude oil and products which could be imported into the United States.[4] The program eventually became quite complex as provisions were made to meet the needs of specific importers and provide competitive assistance to small refiners.

---

[3] 49 Stat. 30 (1935).
[4] Proclamation 4210, 38 Fed. Reg. 9645 (1973).

The effect of the MOIP was to insulate the price of American crude oil from lower world prices. Since there was a relatively fixed ceiling on imports of crude oil, the marginal barrel of oil needed to supply domestic demand would have to come from domestic sources. Thus, the price of domestically produced petroleum set the price charged domestic consumers.

Today, the ability of prorationing and the MOIP to enhance domestic prices has ended. Domestic production of crude oil peaked in 1970. In 1972, the conservation agencies of the producing states set the level of allowables under prorationing at 100 percent. Although the state prorationing programs remain in effect, the setting of the 100 percent level has now become a routine operation. Consequently, there is no longer any "spare capacity" to come to the aid of domestic refiners, and domestic producers produce as much petroleum as is profitable under the current regulatory framework.

In 1973, the MOIP was revised to eliminate volumetric restrictions on the importation of crude. Under the revised program, fees of twenty-one cents per barrel for crude oil, and sixty-three cents per barrel for product are charged for the importation of those commodities. In order to prevent any adverse effects on those who had been receiving products under the old quota system, importers of petroleum under the old program are permitted to import without the payment of license fees those amounts which they were allocated under the quota system. These "fee free" licenses will be phased out by 1980. In essence, the domestic and international crude oil markets would be fully integrated but for the effects of regulations which have arisen since 1971.

The Genesis of Current FEA Regulations. Four significant events since 1971 transformed and dramatically expanded government regulations of the petroleum industry. These events were:

- The presidential announcement in August 1971 of a new economic program which imposed a wage and price freeze, followed by extensive wage and price controls on most sectors of the economy;

- Crude oil and refinery capacity shortages during 1972 and 1973, with resulting pressure on independent refiners and marketers;

- The embargo imposed, following the October 1973 Yom Kippur War, by the Organization of Arab Petroleum Exporting Countries (OAPEC); and

- The establishment and maintenance of high cartel prices for petroleum by OPEC.

Regulations prior to the 1973 embargo. Regulation of the petroleum industry prior to the 1973 embargo occurred primarily as part of the overall Economic Stabilization Program. During 1973, however, attention was also increasingly given to problems arising from various crude oil and product shortages.

On August 15, 1971, President Richard M. Nixon froze all prices and wages except for the prices involving the first sale of agricultural and imported products. The freeze, Phase I of the Economic Stabilization Program, ended on November 13, 1971, and was followed by Phase II of the program which was designed to hold average price increases to no more than 3 percent annually throughout the economy. Up to early 1973, the petroleum industry was not viewed as a special problem by those managing the Economic Stabilization Program, and the industry was governed in approximately the same manner as the other sectors of the economy.[5] During 1972, and continuing into 1973, shortages of crude oil and certain refined products began to develop. Shortages of some products, such as number 2 home heating oil, were in part caused by the price regulations themselves, which set the base price for petroleum products on the basis of their August 1971 levels. During the summer months, gasoline prices are at their seasonal high, and heating oil prices at their seasonal low. Because they could not follow seasonal pricing patterns and raise heating oil prices as winter approached, refiners had no incentive to increase their output of heating oil during the fall months. Thus, as the winter of 1972–1973 approached, regional shortages began to develop. Additionally, the fact that winter was rather cold that year led to localized shortages of propane and to natural gas curtailments. These curtailments, in turn, increased the demand for distillate as consumers who could switch fuels attempted to do so. Although most demands for petroleum products were met, increased pressure was placed on prices.

On January 11, 1973, Phase III came into effect. Under this program, businessmen were to comply on a voluntary and self-administered basis with the standards for cost increases contained in Phase II regulations. As Phase III began, the pent-up pressure on petroleum products in short supply soon led to increased prices, particularly with respect to propane and to home heating oil in New

[5] Charles Owens, "History of Petroleum Price Controls," in *Department of the Treasury Historical Working Papers on the Economic Stabilization Program*, vol. 2 (Washington, D.C.: Department of the Treasury, 1974), p. 1235.

England. Moreover, decreased domestic production of crude oil and a worldwide shortage of crude were raising prices of that input. The prospect of a shortage of domestic refining capacity, particularly as the summer gasoline peak season approached, also foreshadowed further sharp petroleum price increases.

After holding hearings to deal with these problems, the Cost of Living Council (CLC), which was then supervising the Economic Stabilization Program, issued Special Rule No. 1 which reimposed mandatory controls on the major companies in the petroleum industry. These twenty-four companies, each having more than $250 million in sales, did more than 95 percent of the gross industry business.[6] Special Rule No. 1 was to fail as an inflation-fighting tool. This was so because in a shortage situation the noncontrolled suppliers of petroleum were able to increase sharply the prices of their products. The twenty-four major firms were placed under increasing competitive pressure, as the noncontrolled firms bid up the prices of crude oil and products sold by other noncontrolled firms, thus obtaining supplies; the controlled firms were unable to match their bids since they would be unable to pass on increased costs. Most importantly, the regulations did not apply to the many thousands of independent marketers of petroleum products. These dealers, of course, set the prices to be charged to many consumers. These independent dealers were put in the very profitable position of purchasing products at controlled prices from major refiners and reselling them without being subject to CLC price limitations.

The Phase III program terminated on June 13, 1973, with the institution by the President of a sixty-day price freeze. During the freeze, the CLC staff had the opportunity to rethink the entire energy program. In August 1973, Phase IV regulations were announced.[7] As modified in September and October, these regulations were to form the basis for the price controls which were to be adopted by the FEA. The drafters of the Phase IV regulations were keenly aware of the failure of the Phase III system, with its reliance on controls of only the major twenty-four petroleum firms. Therefore, they designed a comprehensive set of regulations to govern the pricing of petroleum and its products by each segment of the industry: producers, refiners, resellers, and retailers.

Perhaps the most important innovation made by the Phase IV regulations was the establishment of a "two tier" pricing system for crude oil. The CLC wished to encourage increased exploration and

---

[6] Ibid., p. 1242.

[7] 38 Fed. Reg. 22536 (1973).

production of additional petroleum, but also did not wish to see the price of existing petroleum production increased. The solution was to establish a system under which an amount of production equal to the 1972 production levels of a given producing "property" was fixed at the May 15, 1973, price for crude oil from that field, plus thirty-five cents per barrel. This was termed *old* oil. Additional oil produced above that amount or oil from properties which did not produce petroleum in 1972 was termed *new* oil, and it was not subject to price control. In order to encourage producers to increase production on existing property, the regulations also provided for a category of *released* oil. This category of crude oil was exempted from price controls and was equivalent to the amount by which current production exceeded the 1972 base level. That is, for each barrel of new crude oil produced, a barrel of old oil would be freed from controls.

After initial amendments, the regulations for refiners and resellers established May 15, 1973, as the base period for prices charged under the program. Refiners were permitted to increase their prices above this level to reflect increases in the cost of the petroleum they purchased and to reflect increased nonproduct costs. Refiners were permitted to "bank" increased petroleum costs to justify future price increases if these costs could not be recovered in the months in which they were incurred. Nonproduct costs could be passed through, subject to the normal CLC "prenotification" procedures and subject to a profit margin limitation. Refiners were permitted to allocate the increased costs attributable to one product to other products, except for three "special products": motor gasoline, home heating oil, and diesel fuel. Costs could only be allocated to these products in a manner directly proportional to the volume of such products sold by the refiner. Additionally, retail price ceilings were set for special products.

Except for those special products with a price ceiling at the retail level, resellers and retailers could charge the purchase price of petroleum products, plus their May 15, 1973, markup. The use of a retail price freeze was implemented on the assumption that it would hold down price increases at the refiner and wholesaler level.[8] Unfortunately, this concept did not work because the CLC staff underestimated the amount of the price increases which refiners would have to absorb and the refiners' lack of concern about the maintenance of dealer margins. Dealer pressure resulted in congressional passage in September 1973 of an amendment to an appropriations bill requiring that FEA permit all segments of the petroleum industry to increase

---

[8] Owens, "History of Petroleum Price Controls," p. 1284.

prices on a dollar-for-dollar basis to cover increased petroleum costs. Regulations implementing the dollar-for-dollar passthrough were promulgated in October. The automatic product-cost passthrough rules were also written into the petroleum allocation act passed in the embargo period.

The increasingly short supply of crude in certain petroleum products during the spring of 1973 led to increasing federal involvement in managing the allocation of petroleum as well as setting its prices. The shortage led to increasing pressure on many independent refiners and marketers who were unable to purchase crude oil and products. Responding to pressure from the independent sector, Congress approved amendments to the Economic Stabilization Act of 1970, effective April 30, 1973, giving the President the authority to allocate crude oil and petroleum products.

On May 21, 1973, the Nixon administration announced a voluntary petroleum allocation program for both crude oil and products.[9] The program was administered by the Office of Oil and Gas in the Department of the Interior. Compliance with the program was mixed.

On June 29, 1973, the President established an Energy Policy Office (EPO) which, on August 13, 1973, proposed regulations for the mandatory allocation of crude oil, petroleum products, and liquefied natural gases.[10] The proposed comprehensive allocation programs were not adopted. Rather, regulations were adopted in October 1973 dealing with the allocation of middle distillate fuels and propane.[11]

The embargo. Following the October 1973 Yom Kippur War, and the imposition of an embargo on the sale of crude oil and petroleum products to the United States by OAPEC, the first concern of the government became the allocation of petroleum, as the prospect of severe shortfalls loomed. On November 1, 1973, an Office of Petroleum Allocation was established in the Department of the Interior to administer the voluntary program and the mandatory propane and middle distillate regulations. On November 27, 1973, Congress passed the Emergency Petroleum Allocation Act of 1973 (EPAA).[12] This legislation required the President to promulgate mandatory regulations for the allocation of crude and petroleum products, and for the regulation of their prices. On December 4, 1973, the President by executive order established the Federal Energy Office (FEO). Proposed alloca-

---

[9] 38 Fed. Reg. 13588 (1973).

[10] 38 Fed. Reg. 21797 (1973).

[11] 38 Fed. Reg. 27397, 28660 (1973).

[12] Emergency Petroleum Allocation Act, P.L. 93-159 (1973), as amended, hereinafter EPAA.

9

tion regulations were first published for comment on December 12, 1973. The final allocation regulations, effective immediately, were published on January 15, 1974.

These regulations, which were subject to several technical amendments shortly after their publication, formed the basis for the allocation regulations in effect today. Essentially, the regulations froze supplier/purchaser relationships between crude producers, refiners, wholesale resellers of petroleum, wholesale consumers, and certain other end users as of a base date or with reference to a base period. Notably excluded from this fixed-supply scheme was the ordinary motorist. Base period volumes were established, generally reflecting a period within 1972. Suppliers were obligated to supply, but purchasers were not obligated to purchase, these base period volumes. Next, a system of priorities was established among consumers. The volumes to be allocated under these priorities were subject to pro rata reduction by a supplier based on the supplier's "allocation fraction." This figure was a percentage determined by the ratio of the supplier's available product to his supply obligation. Only sales to the Department of Defense, for certain agricultural uses, and for certain medical space-heating requirements were exempt from reduction under the allocation fraction.

A special program, called the "buy/sell program," was established to equalize the quantity of crude oil available to each refiner. Under this program, refiners having a higher percentage of crude oil supplies available, in relation to their refining capacity, than the national average were required to resell their crude to refiners with below normal crude availability.

Under the allocation scheme, FEO became involved in supervising changes in ongoing business relationships, since FEO approval or notification was required for the termination or establishment of changes between purchasers and suppliers covered by the program. Additionally, FEO would help new customers find suppliers if the prospective purchaser were unable to do so.

During the embargo period, price regulations remained substantially unchanged. The only major modification was the establishment of a program to encourage the production of middle distillate by permitting refiners to increase distillate prices when the percentage of the refiner's output devoted to middle distillate also increased. The incentive mechanism was revoked on February 27, 1974. Legislation approved November 16, 1973, exempted from crude price regulations crude oil produced from "stripper well" leases, that is, those properties which produced an average of less than ten barrels of petroleum per well per day.[13] This provision was subsequently incorporated into the EPAA. Additionally, in mid-December 1973, the CLC raised the price of old oil one dollar per barrel, in response to the growing difference between new and old oil which resulted from the sharply increased price of world petroleum.

With the formation of the FEO in December 1973, responsibility for price and allocation regulations was placed under the same organizational roof for the first time. When the mandatory allocation regulations were published on January 15, 1974, they were published in a unified manner as part of a newly created title 10, chapter II, of the *Code of Federal Regulations*.

Regulatory developments from the end of the embargo to the passage of the 1975 Energy Policy and Conservation Act. After the end of the embargo in 1974, the development of petroleum regulations was subject to two conflicting forces. The first was the desire of many to eliminate regulations entirely as adequate supplies of crude oil and products were reestablished. The second was the strong sentiment for continued regulation of oil coupled with the recognition that, if the regulations were going to be retained for any appreciable length of time, they would have to be revised to accommodate the dynamics of the petroleum industry.

On May 27, 1974, legislation was aproved establishing the Federal Energy Administration. This statute, the Federal Energy Administration Act of 1974 (FEAA),[14] did not make substantive changes in the regulation of the petroleum industry. Rather, it simply established the structural and procedural basis for FEA and transferred various energy functions to the agency. Thus, effective June 27, 1974, implementation of the various statutes authorizing regulation of the petroleum industry was, for the first time, placed on a firm, unified statutory foundation.

From May 1974 through 1975, many modifications, most of them relatively minor, were made to the regulations. The two major changes concerned the buy/sell program and the establishment of crude oil "entitlements." On May 14, 1974, the buy/sell program was revised to limit the class of sellers to the fifteen largest integrated refiners, with only the small refiners defined as eligible buyers. The previous program had generated resentment among the majors when they were required to supply crude petroleum to each other. Also, the program

---

[13] Trans-Alaska Pipeline Authorization Act, P.L. 93-153, section 406 (1973).

[14] Federal Energy Administration Act, P.L. 93-275 (1974), as amended, hereinafter FEAA.

was revised to remove incentives against the purchase of additional imported crude.[15]

On December 4, 1974, FEA established the entitlements program, which attempted to equalize the cost of crude oil among every refiner.[16] When the two-tier pricing system was set up in August 1973, the extraordinary increases in the world price of crude oil as the result of OPEC action had not yet occurred. At the time, it was contemplated that the difference between upper and lower tier crude would probably be in the vicinity of one dollar per barrel. During 1974, however, the differential increased to about eight dollars per barrel. Consequently, the average price that refiners had to pay for their crude varied greatly, depending on the proportion of lower and upper tier crude oil used as inputs. Those refiners which had a greater than average proportion of high-cost upper tier oil were placed at a competitive disadvantage.

Technically an allocation program governing old oil, the entitlements program mandated that refiners were "entitled" to use only the average national percentage of old oil as an input. Those having more old oil than the national average were to purchase "entitlements" to use their excess lower tier oil, while those having less than normal supplies of low-cost crude oil would sell their unused entitlements. The FEA was to act as a clearinghouse for these transactions. Small refiners were also given additional entitlements as a subsidy to help them compete against the majors. This "small refiner bias" meant that small refiners having higher than normal supplies of old oil had to pay proportionately less than majors for new oil, or they had extra entitlements which they could sell.

Other modifications during this period included elimination of the prenotification requirement with respect to the passthrough of in-increased nonproduct costs and an identification of certain specific nonproduct costs which refiners were able automatically to pass through on a dollar-for-dollar basis,[17] and the elimination of gasoline from the "special products" category.[18]

Current regulation of the industry. In December 1975, the federal regulatory structure was given more firm direction with the passage of the Energy Policy and Conservation Act (EPCA).[19] The

---

[15] 39 Fed. Reg. 17287 (1974).

[16] 39 Fed. Reg. 42246 (1974).

[17] 39 Fed. Reg. 42368 (1974).

[18] 39 Fed. Reg. 39259 (1974).

[19] Energy Policy and Conservation Act, P.L. 94-163 (1975), as amended, hereinafter EPCA.

statute provided for the phasing out of price regulations on crude oil over a forty-month period. It established a fixed national average price for all crude oil which is permitted to escalate gradually over that time. FEA was given discretion to establish various price tiers to achieve the national average price level. Utilizing this authority, FEA retained the same price level for old oil, initially estimated to be $5.25 per barrel, and established an upper tier price for what was previously new, released, and stripper-well production, initially estimated to be $11.28 per barrel. Imported oil, which was not under control, constituted a third tier. FEA was given the discretion to increase the average price for all crude, initially set at $7.66 per barrel, up to 10 percent per year. Price increases beyond this level were to be submitted to Congress for possible veto.

EPCA recognized that the shortage conditions which had brought about controls over refined products were no longer present. Therefore, FEA was permitted to begin a program of phased decontrol of refined products if it could make several findings relating to the sufficiency of supply of the product and reduce chances for adverse impact on inflation and on competition within the petroleum industry. Such decontrol proposals were subject to veto by either house of Congress.

Utilizing this authority, FEA has proposed, and Congress has acquiesced in, the decontrol of a large number of petroleum products. These include residual fuel, middle distillates, naphthas and gas oils, various specialty products, and naphtha jet fuel. The only products currently remaining under control are motor gasoline, aviation gasoline, kerosene-based jet fuel, natural gas liquids, natural gasoline, butane, and propane. The effect of decontrol is to exempt the product from the price and allocation regulations subject to the reimposition of controls by the President if the need arises. The controls may subsequently be removed without congressional approval.

Although middle distillate has been decontrolled, a Special Rule No. 3 to the allocation regulations has been established to provide an emergency set-aside of that product in case specific areas or dealers face a shortage of that product. Also, a monitoring program exists with respect to middle distillate prices, and FEA is committed to institute appropriate regulatory action if middle distillate prices exceed a specified level.

Following the passage of EPCA, FEA continued to make changes in its regulations. In revising the crude oil pricing regulations to conform with the act, it eliminated the released oil category and instead adopted a mechanism whereby the base period production level for a given field would be placed on a declining basis to correspond to the field's historical decline rate.[20] Additionally, the definition of *property* was revised, effective September 1, 1976, to permit a producer in certain circumstances to treat oil from a newly developed reservoir as oil from a new property even if the reservoir was located on the same lease.[21] Regulations governing banks were revised to incorporate limitations required by EPCA, and the ability of refiners and resellers to bank nonproduct costs was established.[22]

On August 14, 1976, the Energy Conservation and Production Act (ECPA) was passed.[23] The statute again exempted stripper oil (recontrolled by EPCA) from price controls and mandated that, for the purpose of computing the national average price, stripper production be treated as having been sold at an imputed price. This action was designed to prevent the exemption of stripper-well crude oil from having an effect on the price of crude oil remaining subject to controls. The act also required FEA to establish greater price incentives for the use of tertiary production methods. It should be noted that for individual cases, exceptions to these regulations are authorized by decisions made under the exception process.

The Structure of FEA Regulations

This section will attempt to analyze the FEA regulatory structure on the basis of the "programs" that are inherent in it (except for the Mandatory Oil Import Program) rather than on the basis of a simple price or allocation dichotomy. Three major categories have been identified: control of crude oil prices and maintenance of product prices below world product prices, including control of relative prices to consumers of various petroleum products; preservation of the "independent" and small business sector of the petroleum industry and the preservation of competition generally (not necessarily identical goals); and allocation of scarce supplies. Additionally, two other programs appear to exist within regulations designed to achieve the other three goals: promotion of crude oil production; and the promotion of the efficient use of petroleum products and efficiency in the operation of the petroleum industry. It is readily apparent that these programs track the statutory goals set by Congress.

In implementing those goals, FEA was subject to several statutory mandates governing the content of the regulations, the factors to be

---

[20] 41 Fed. Reg. 4931, 15566 (1976).

[21] 41 Fed. Reg. 36172 (1976).

[22] 41 Fed. Reg. 15330 (1976).

[23] Energy Conservation and Production Act, P.L. 94-385 (1976), hereinafter ECPA.

14

considered in the regulatory decision-making process, and the procedures for amending the regulations. Most of these broadly define the outer boundaries of FEA discretion rather than specifically prescribe precise substantive provisions to be implemented by the agency.

The statutory provisions which require that FEA regulations contain certain substantive provisions are few. First, the price regulations must provide for a dollar-for-dollar passthrough of net increases and decreases in the cost of crude oil, residual fuels, and products at all levels of distribution.[24] Second, the price regulations must use the same base date for computation of markup, margin, and posted price for all marketers and distributors of crude oil, residual fuels, and products at all levels of distribution.[25] Third, the crude oil price regulations must establish ceiling prices for the first sale of domestic crude oil which will result in an actual weighted average first-sale price of $7.66 per barrel, except as adjusted upward by specified procedures.[26]

A secondary category of mandate, placing less of a constraint on the exercise of the agency's discretion, requires that the regulations achieve a certain effect if that effect is consistent with the achievement of other objectives. Thus, section 4(c) of EPAA requires that the regulations, "[t]o the extent practicable and consistent with" the statutory objectives, allocate supplies of crude oil and product to the independent sector of the industry in amounts corresponding to that sold or supplied to independents during 1972, and reduce those supplies on a pro rata basis in the event of a decrease in supply.[27]

A similar approach is found in the section of EPAA requiring that the allocation program, to the extent practicable and consistent with the overall objectives, not deny liquified petroleum gas (LPG) to industrial users without adequate alternatives.[28]

A third type of statutory constraint on FEA's regulatory development program structures the decision-making process, but does not necessarily control the content of the regulations. Statutory provisions require that FEA, before taking certain actions, make specified findings. This approach is most apparent in FEA's decontrol authority,[29] but also applies to limitations on cost banks.[30]

---

[24] EPAA, sections 4(b)(2)(A) and 9.

[25] Ibid., section 4(b)(2)(C).

[26] Ibid., section 8(a).

[27] Ibid., section 4(c)(1).

[28] Ibid., section 4(c)(5).

[29] Ibid., section 12(b).

[30] Ibid., section 4(b)(2).

Furthermore, some regulatory actions, again primarily with respect to decontrol, require that FEA submit its proposed actions to Congress for approval and support the proposal with specified data and findings.[31]

For the most part, however, FEA's discretion with respect to the content of the regulations is very broad, consistent with the intent of Congress "to hew a sphere of responsibility"[32] within which FEA might construct the program. The agency is charged with shaping a program which must serve, to the maximum extent practicable, the following nine statutory objectives:

(A) protection of public health (including the production of pharmaceuticals), safety and welfare (including maintenance of residential heating, such as individual homes, apartments and similar occupied dwelling units), and the national defense;

(B) maintenance of all public services (including facilities and services provided by municipally, cooperatively, or investor owned utilities or by any state or local government or authority, and including transportation facilities and services which serve the public at large);

(C) maintenance of agricultural operations, including farming, ranching, dairy, and fishing activities, and services directly related thereto;

(D) preservation of an economically sound and competitive petroleum industry, including the priority needs to restore and foster competition in the producing, refining, distribution, marketing, and petrochemical sectors of such industry, and to preserve the competitive viability of independent refiners, small refiners, nonbranded independent marketers, and branded independent marketers;

(E) the allocation of suitable types, grades, and quality of crude oil to refineries in the United States to permit such refineries to operate at full capacity;

(F) equitable distribution of crude oil, residual fuel oil, and refined petroleum products at equitable prices among all regions and areas of the United States and sectors of the petroleum industry, including independent refiners, small refiners, nonbranded independent marketers, branded independent marketers, and among all users;

(G) allocation of residual fuel oil and refined petroleum products in such amounts and in such manner as may be

---

[31] Ibid., sections 12(c) and (d).

[32] Consumers Union v. Sawhill, 525 F.2d 1068 (Temporary Emergency Court of Appeals 1975).

16

necessary for the maintenance of, exploration for, and production or extraction of—

    (i) fuels, and

    (ii) minerals essential to the requirements of the United States, and for required transportation related thereto;

    (H) economic efficiency; and

    (I) minimization of economic distortion, inflexibility, and unnecessary interference with market mechanisms.[33]

Congress fully realized that these goals were broadly drawn and inevitably contradictory.

> The listing of objectives in successive paragraphs (A) through (I) in section 4(b)(1) is not intended to establish any order of priority. These are collective goals, and the conferees have not attempted to discern an order of precedence or value one against another. It is fully recognized that, in some instances, it may be impossible to satisfy one objective without sacrificing the accomplishment of another.[34]

Courts have recognized that, given the nature of these objectives, the agency must be relatively free from legal challenge on the ground that a particular regulation serves one goal necessarily at the expense of another. FEA, then, has the responsibility to balance, in its informed discretion, statutory energy objectives and to fashion, subject to a few specific statutory mandates, the nation's energy program.

Preventing Crude and Product Prices from Reflecting the OPEC Price of Crude. Perhaps the keystone of the current regulatory program is the congressionally required control of crude oil prices and the complementary regulations for the control of petroleum product prices. The program can best be understood in terms of the desire of Congress (both express and implicit) to prevent the high OPEC prices for crude oil and the concomitant high world prices for products from setting domestic prices. The two primary adverse effects of permitting such price setting by OPEC were perceived as the reaping of windfall gains by owners of domestic crude and the recessionary effects of increased petroleum prices on the economy.[35] Section 8(a) of the EPAA, as added by the EPCA and amended by the ECPA, requires FEA to establish a pricing structure which would result in the average

---

[33] EPAA, section 4(b)(1).

[34] H.R. Rep. No. 93-628, 93rd Congress, 1st session (1974), Conference Report.

[35] See H.R. Rep. No. 94-340, 94th Congress, 1st session (1975), pp. 7-9.

17

national price for domestically produced crude to be set initially at $7.66 per barrel. This price could be adjusted upward at a rate of up to 10 percent annually if necessary to increase the incentive for production and to compensate for the effects of inflation. Price increases above that level may only be authorized when the FEA administrator makes certain express findings, and Congress does not veto the proposals.

The crude pricing regulations are today essentially "rent controls," in that they are intended to prevent owners of lower cost old petroleum from gaining wealth as the market price of crude increases toward the world price of crude set by OPEC, while their own production costs remain static. At the same time, producers of new crude are permitted to receive prices approximating the world price of petroleum so that there will be an incentive to bring on line petroleum whose cost of production approaches the world selling price. This rationale is to be distinguished from the earlier purely anti-inflationary goals of the Economic Stabilization Program. Its rules were aimed at holding down prices and making businesses absorb increasing costs.

The control program also attempts to prevent product prices from rising to world levels which are based on the cost of OPEC crude, a situation which would prevail if domestic product shortages were to develop and product imports were the marginal source of supply. In such cases, middlemen would be able to enrich themselves by charging the economic "rents" which domestic crude producers could not charge due to the crude oil pricing regulations. The product regulations are also outgrowths of the Phase IV inflation control program.

Implementing its statutory mandate, the FEA has established a three-tier system of crude pricing.[36] The first tier is composed of "old" oil, that is, the average monthly amount of old oil produced by a "property" during 1975 (1972 may also be used as a base period). Such old oil may currently be priced at no higher than the May 15, 1973, posted price for such oil, plus $1.48 per barrel, an average of $5.16 per barrel. The second tier is composed of "new" and "released" oil, oil in excess of that produced in the base year. Such oil is currently to be priced at the September 30, 1975, posted price for that crude, less $1.05 per barrel, an average of $11.65 per barrel. The third tier is composed of imported oil, the first-sale price of which is not controlled, and now under ECPA also includes oil from stripper-well properties, that is, properties which produced an average of less than ten barrels per well per day during a preceding consecutive twelve-month period.

---

[36] 10 C.F.R., sections 212.72-77.

This pricing mechanism is similar to that adopted by the August 1973, Phase IV rules. At that time, there were only two tiers: controlled old oil and uncontrolled new and released oil, which it was assumed would reach the import price. The multitiered system was designed to stabilize the price of crude produced from existing properties while providing an incentive to producers to search out higher cost production, either from new properties or through the use of enhanced recovery or other techniques to increase production from old properties.

FEA price regulations set the price which a refiner or reseller may charge for a product at the product's May 15, 1973, price plus the seller's allowable cost increases. In setting out these allowable costs, the rules continue the Phase IV distinction between product and nonproduct (operating) costs. Currently, both refiners and sellers are permitted to make dollar-for-dollar passthroughs of crude and purchased-product costs.[37] The refiner regulations are quite complex and take into account the various ways a refiner may purchase crude and then resell it; for example, as product, as a voluntary resale, as a resale under the allocation program, or as a resale through the FEA buy/sell program.[38] Additionally, as will be explained below, costs associated with the crude purchased for the production of various products may be reallocated to other products.

Nonproduct costs are regulated in a manner more in keeping with the Economic Stabilization Program than with the current "rent control" situation. Initially, refiners were permitted to pass through all nonproduct costs which were not disallowed after "prenotification," as had been required by the Cost of Living Council, before they could increase prices.[39] This requirement was removed in late 1974. Instead, nonproduct costs could automatically be passed through, but they were limited to several specified categories: refinery fuel, labor costs, additive costs, utility costs, pollution-control costs, interest costs, marketing costs, and container costs. Excluded from this list were increased capital expense costs, creating a disincentive to refinery investment. In late 1976, FEA was proposing to include as permissible nonproduct cost increases: depreciation, maintenance, and

[37] 10 C.F.R., sections 212.83 and 212.93.

[38] Additional complex regulations in 10 C.F.R., section 212, subpart K, govern the pricing of natural gas liquids. These regulations must take into account processing shrinkage and the relationship between natural gas and petroleum-based natural gas liquid.

[39] For a general history of refinery price regulations, see 41 Fed. Reg. 31863-69 (1976).

taxes (other than income taxes).[40] A further requirement that an increase in income from the passthrough of nonproduct costs not increase the refiner's profit beyond his base period margin was removed in early 1976. Refiners may also vary their gasoline prices by region, subject to a three-cents-per-gallon limitation.

Resellers of petroleum products, however, are still subject to a "cents per gallon" limitation on the total amount of nonproduct costs which they may pass through. Currently, gasoline retailers may not increase their prices to reflect increased nonproduct costs in an amount greater than three cents per gallon over their May 1973 gross margins (except in Alaska, where the margin is five cents). In a statement defending the resellers' cents-per-gallon limitation on November 25, 1975, FEA stated that it was not obligated to grant a dollar-for-dollar passthrough of nonproduct costs by the current statute and cited as precedent for failure to do so the Economic Stabilization Program and Phase IV regulations.[41] FEA gave two basic reasons why such a restriction should continue. First, it was thought unreasonable to expect the hundreds of thousands of small retail outlets to calculate successfully their per-gallon nonproduct cost. In such a case, administrative considerations would require that industry-wide guidelines be used where accurate firm-by-firm data was not available for compliance purposes. Second, for those firms that could successfully make such calculation, the limitation has provided a ceiling on the passthrough of nonproduct costs. Such a ceiling would benefit the consumer by requiring suppliers to absorb cost increases. As the FEA concluded:

> In view of the administrative need for standard markups and the functions they are expected to serve, and in order to assure equitable prices to consumers by minimizing the inflationary impact of price increases to the maximum extent possible, FEA believes that the maximum markup selected in each case should be determined according to conservative criteria of measurement, so that the nonproduct cost increases of the majority of the dealers concerned equal or exceed that level.[42]

It should be noted that, even when sellers are permitted full dollar-for-dollar passthroughs of both product and nonproduct costs, their net margins become fixed on a cents-per-gallon basis and become a

---

[40] Ibid.

[41] 40 Fed. Reg. 54561 (1975).

[42] Ibid., p. 54564.

decreasing percentage of gross sales, and they are worth less in real terms as a result of general inflation.

Because of seasonal or other marketing factors, refiners and dealers may find that they are unable to recoup cost increases that they would otherwise be permitted to pass on to their customers. Therefore, sellers may "bank" such costs to be passed on when market conditions permit. Were this not to be the case, sellers either would find themselves undergoing severe hardship (assuming the FEA price limits were below the market price) since they could not recover their increased costs or, conversely, would be forced to increase prices against normal market conditions, just to establish a higher price for later sales. Although product costs could be banked under the original Phase IV regulations, nonproduct costs could not be banked until February 1, 1976.

In passing EPCA, Congress established certain restrictions on the ability of a refiner to justify an increase in his prices by referring to his bank of unrecovered costs.[43] Under regulations implementing these limitations, costs incurred after February 1, 1976, generally can be passed through in the two months following the month in which they were incurred.[44] If not passed through during that time, they are placed in a cost bank. This bank may be drawn down at the rate of 10 percent of the total bank in any given month. Costs incurred prior to February 1, 1976, are considered a separate bank and may also be drawn down, subject to a 10 percent limitation. Within these categories, there are further cost hierarchies involving crude costs, purchased-product costs, and nonproduct costs. In its April 12, 1976, decision implementing the nonproduct banking regulations, FEA recognized that, in the current environment of product surplus, the limitation on the draw-down of banks was unnecessary. Rather, it saw such rules as a precaution against future shortages:

> Although FEA does not consider such limitations to be necessary under current market conditions in which adequate supplies are exerting downward pressure on prices, it recognized the desirability of providing a means in price regulations that would prevent sharp price increases from occurring due to the possible application of excessive amounts of unrecognized costs in the event a period of supply shortages were to occur.[45]

---

[43] EPAA, section 4(b)(2)(B), as amended by EPCA.
[44] 10 C.F.R., section 212.83(f).
[45] 10 C.F.R., section 212.93(e).

Resellers of petroleum products have similar, but simpler, banking provisions.

It should be noted that the "order of recoupment" provisions of the refiner regulations controlling the order in which the various types of banks may be drawn down are quite complex. FEA has recently discovered that many refiners, possibly relying on the interpretation of these rules in FEA's audit manual, may have unlawfully treated as much as $1.3 billion to be eligible for passthrough. The extent to which any portion of this amount may actually have been reflected in prices charged has not been determined, however. A 1976 rulemaking was to deal with this problem.[46]

An additional element of FEA price regulations are the provisions which control the relative prices of different petroleum products. FEA regulations permitted refiners—and, since May 1976, distributors—to reallocate increased costs attributable to one petroleum product to the price charged for another. Such a system gave sellers added flexibility in pricing their products according to market demand while restraining the total revenue that could be received from the sale of petroleum products. The reallocation regulations had their genesis in the Phase IV special products rules. "Special products" were limited to motor gasoline, diesel fuel, and home heating oil. Only a proportionate share, based on sales volumes, of increased crude costs could be allocated to special products, while refiners were free to allocate any increased crude costs to other products.

Prior to the exemption of various products in 1976, FEA regulations reflected the EPCA requirement that no more than a proportionate distribution of crude costs could be made for the following products: number 2 home heating oil and diesel fuel, aviation fuel of the kerosene or naphtha type, and propane produced from crude oil. These requirements could be waived by FEA without congressional concurrence upon a showing that refinery operations justify a deviation from the rule, that the deviation is consistent with the general purposes of the EPAA, and that the change would not result in "inequitable prices" for any class of users.[47]

Under the regulations prior to the various exemptions, crude costs were to be allocated on not more than a proportionate basis to each of the products specified by the statute and to a class of "general refinery products." These general refinery products included residual fuel oil, greases, lubricants, naphthas, gas oils, aviation gasoline, number 1 heating oil and number 1-D diesel fuel, natural gas liquids,

---

[46] 41 Fed. Reg. 43953 (1976).
[47] EPAA, section 4(b)(2)(D).

butanes, natural gasoline, and miscellaneous other refinery products. Even though general refinery products as a whole could receive only a proportionate share of increased crude and product costs, a refiner was free to allocate crude costs within the class. All remaining costs not recouped with respect to sales of the three special classes or the class of general refinery products could be allocated to gasoline.[48]

Thus, in contrast to the initial program, gasoline was singled out as the product on which unrecovered costs from the refining of all other products could be placed. In announcing the elimination of gasoline from the special products rule, the FEA stated that it did not expect significant price increases but that, to the extent that such increases did occur, "they could be expected to result in further reductions in consumption." This was so because FEA concluded that gasoline consumption was more price elastic than previously thought. Also, to the extent that costs could now be shifted to gasoline, pressure to increase prices on other products in the general category would be reduced:

> Since gasoline accounts for a relatively large volume of the production of most refiners, even a slight increase in gasoline prices could contribute significantly to reducing prices for other products which are not in the special products category.[49]

A special rule with respect to unleaded gasoline controls the relationship between unleaded gasoline and regular gasoline prices in instances where the refiner or reseller of that product was not selling unleaded in the price base period.[50]

The rule generally limits the refiner to charging one cent per gallon over the price of a gallon of regular gasoline of comparable octane and prevents the retailer from charging a greater markup than is permitted for other products. In making its calculations of price, FEA left out any considerations of supply, stating simply that according to an Environmental Protection Agency (EPA) study "unleaded refining costs are only slightly above leaded refining costs" and that therefore a mere one-cent-per-gallon surcharge would be appropriate. Such questions as capital cost were left for some as yet unfinished proceeding.[51] With respect to marketers, "FEA has tentatively concluded that price increases previously authorized to take

---

[48] 10 C.F.R., section 212.83(d).
[49] 39 Fed. Reg. 39259 (1974).
[50] 10 C.F.R., section 212.112.
[51] 39 Fed. Reg. 24923 (1974).

into account increased nonproduct costs should, in most cases, be adequate to cover the increased nonproduct cost of handling unleaded gasoline."[52]

The FEA has also concerned itself as to whether service station operators should be permitted to charge gasoline customers for increased operating costs not directly related to the sale of gasoline. In adopting a rule of thumb that only 80 percent of increased gas station operating expenses should be attributable to gasoline—the other 20 percent to other service station functions—the FEA necessarily controlled the relative price of gasoline and other service station products:

> Similarly, it is appropriate for FEA to attempt, if possible, to segregate those operating cost increases which are properly allocable to gasoline sales from those which are allocable to sales of tires, batteries, and repair services, for example, in order to provide a guideline for determining the extent to which increases in total operating costs should be permitted to be passed through in the form of price increases on gasoline and the extent to which increases in total operating costs are or should be passed through on tires, batteries, etc. In other words, FEA believes that some operating cost increases are or should be recouped on sales of non-gasoline items, and that therefore only operating cost increases which have been "discounted" to reflect an apportionment of such increases to non-gasoline sales should be used to justify the standard maximum markup on gasoline sales.[53]

FEA has also been concerned with relative regional prices. For example, such concerns played an important part in the establishment of the entitlements program for imported residual oil. Initially, sales of residual oil to the East Coast area had provided competitive problems because one refiner, Amerada Hess, had a refinery in the Virgin Islands and was included in the program for crude oil entitlements. Other companies with refineries located in the Caribbean received no entitlements and their prices for residual fuel oil were thus at a competitive disadvantage. In conjunction with the placing of residual fuel controls in standby status, FEA "solved" or attempted to solve the competitive problem by reducing the entitlements which Hess could receive and by establishing entitlements payable to importers of foreign residual oil equal to one-third per barrel of the crude entitlement.

---

[52] Ibid., p. 24924.
[53] 40 Fed. Reg. 54561 (1975).

24

While the initial problem arose out of competitive concerns, the solution adopted was in great part dependent on the question of regional equity:

> FEA is, however, particularly concerned with the regional impact of each of its proposed alternatives. In this regard, FEA concurs with the comments received that the first proposed alternative [eliminating crude entitlements] would result in energy cost increases on the East Coast, whereas the second alternative [utilizing product entitlements] would lower East Coast energy costs and increase those for other regions of the United States. Because of the adverse impact both alternatives have upon different areas of the country, neither alternative has been adopted. Instead, the amendment adopted hereby combines elements of the two alternatives in a manner which is expected to remove the current market distortions but which assures virtually no changes in the regional impact of the amended entitlements program compared with the program prior to this amendment.[54]

Similarly, until their recent amendment, the price regulations required refiners to price product sales to a class of purchasers on a uniform national basis. The regulations now permit gasoline pricing by Petroleum Administration for Defense (PAD) districts, but the districts are limited to a maximum differential of three cents between regions. This limitation was adopted in part to protect independent distributors and in part to ensure that price differentials would not be so great that "consumers in particular regions would be unduly burdened."[55]

Preserving Small and Independent Businesses in the Petroleum Industry, and Promoting Competition. Congress, in establishing the petroleum regulatory program, expressed concern with the fate of the small or independent refiner and the independent marketer and the competitive alternatives they provide to the majors. The broader policy was preservation and encouragement of competition. Protection of competition, however, is not the same thing as protection of individual competitors, since the freezing of market structure reduces the ability of firms to compete for new customers. Although the reconciliation of these goals is difficult, Congress clearly intended that such reconciliation be attempted.

---

[54] 41 Fed. Reg. 13899 (1976).
[55] 41 Fed. Reg. 30022 (1976).

The first element of the regulatory effort to preserve independent business in petroleum industry is the freezing of supplier/purchaser relationships.[56] Pursuant to the supplier/purchaser rule, a supplier must supply his purchasers with their base period volume and, in the case of crude, the supply obligation runs with the purchaser of the producing property. Purchasers of crude or products, however, are not required to purchase any of their allocation.

For example, the initial petroleum allocation regulations effective in January 1974 froze crude supplier and purchaser relationships as they existed on December 1, 1973. This freeze date was later revised to January 1, 1976. According to an affidavit filed in a case challenging certain of the crude oil regulations, an associate assistant administrator of FEA stated that the initial freeze was adopted for three basic reasons: first, to maintain intact most of the crude oil purchaser and seller framework which existed during calendar 1973 and thus minimize the potential for disruption when annual crude contracts were terminated at the beginning of 1974; second, to establish a floor upon which the crude oil allocation program could be built—"without maintaining existing supplier/purchaser relationships, it would have been virtually impossible to make the estimates upon which the 'buy/sell' allocation program depended"; third, to preserve access by independent and small refiners to price-controlled domestic crude and to prevent them from being cut off by the majors.[57]

Crude supplier/purchaser relationships can now be terminated at the option of the purchaser, if all subsequent purchasers of the crude who would be affected by the change consent, or by the supplier, if the purchaser fails within fifteen days to match a bona fide offer of a higher price from another purchaser.[58] Additionally, a producer may terminate sales from a stripper well if the new purchaser is a refiner with less than 175,000 barrels-a-day capacity. It is interesting to note that the proposed rulemaking would have made it possible for a supplier to terminate if another purchaser offered superior services, such as with respect to the gathering of crude. This amendment was rejected on grounds of administrative difficulty because the present purchaser could have no real way of judging offers of better service.[59] The revised purchaser substitution provisions are a clear example of the tension in the program between promoting competition, protecting

---

[56] For example, 10 C.F.R., sections 211.9, 211.63, and 211.105.

[57] Affidavit of John Vernon, Condor Operating Co. v. Sawhill, 3 FEA, par. 26,159 (TECA 1975).

[58] 10 C.F.R., section 211.63(d).

[59] 41 Fed. Reg. 24339 (1975).

26

individual purchasers, and developing an administratively feasible program which is capable of being enforced.

Supplier/purchaser freezes are also in effect with respect to petroleum product wholesalers and retailers. These relationships vary with respect to the manner of termination, for example, whether it must be done mutually or on one party's demand, and whether FEA consent is required. Regulations also provide for the assignment of a supplier and a base period volume by FEA for new entrants into the petroleum distribution industry. A new, low-cost supplier will be assigned to a reseller under the exception process if the reseller's current supplier is charging the reseller prices significantly in excess of those being charged the reseller's competitors by their suppliers.[60] Additionally, the FEA, through the exception process, will assist small businessmen in getting supply contracts with the U.S. government by substituting small businesses for larger firms as the assigned supplier for specified government installations. This program is administered in conjunction with the Small Business Administration.[61]

The crude oil allocation program, also known as the buy/sell program, was also established at the time of the first allocation program and became effective in its present form on June 1, 1974. Its purpose is to correct imbalances in access to crude that existed between the fifteen major integrated refiners and small and independent refiners.[62] The amount of oil which is to change hands is based on a certain percentage of the difference between a refinery's crude supply in 1972 and its crude use in a base period in 1974 representing the embargo period. The allocations are made on a quarterly basis from the fifteen major refiners (called refiner-sellers) to the smaller refiner-buyers. Each refiner-seller must supply a percentage of the total supplier obligations based on aggregate needs of each refiner-buyer. Sales of crude oil to refiner-buyers are at prices representative of the particular seller's weighted average landed cost of imported crude oil.

FEA regulations also deal with competitive imbalances on the refiner level caused by the multitier crude price regulations. With the quadrupling of oil prices by OPEC, the spread between the price levels became quite significant. Not unexpectedly, complaints were heard from many refiners that they were being placed at a competitive disadvantage because competing refiners had access to a significantly greater proportion of old crude at lower prices. In response to these

---

[60] For example, Colonial Oil Co., 2 FEA, par. 83,201 (July 3, 1975).

[61] For example, Grimes Oil Co., 3 FEA, par. 83,046 (December 15, 1975).

[62] 10 C.F.R., section 211.65.

complaints, FEA in November 1974 established the entitlements program to reduce to a competitive range disparities in crude acquisition costs for all domestic refiners so as to place them on roughly equal competitive footing in terms of crude supply cost.[63]

Under the program, FEA determines a national refiner supply/capacity ratio which takes into account the proportion of crude represented by each of the three tiers in the current pricing system. The heart of the program is the concept that one must have an "entitlement" to use lower priced domestic oil. Entitlements are issued to refiners based on their crude runs, adjusted for the amount of crude processed by other refiners under a processing agreement. Those refiners having access to an above-average percentage of lower priced oil must purchase entitlements to use that crude. Likewise, those refiners having a less-than-average supply of old oil can sell their surplus entitlements. Technically, the entitlement program is an allocation scheme, but physical transfer of old crude oil so that all refiners process the national average would be wasteful. FEA, in fact, serves as the clearinghouse in that it calculates the entitlement obligation and directs the transfer of funds. In late 1976, the entitlement price was $7.90 a barrel, and the national annual volume of entitlement sales was over $1 billion.

An important aspect of the entitlements program is the so-called small refiner bias. Under this regulatory program, small refiners are granted additional entitlements, effectively a subsidy to that group. The bias was instituted to compensate small refiners for their lack of the economies of scale and is based on advantages which they previously received under the oil import quota system. Under the import quota system, abolished in April 1973 in favor of a license fee mechanism, all refiners were given "tickets" to import crude oil. At that time, imported crude was less expensive than domestic crude, so that the right to share in the restricted volume of imported trade had economic value. Small refiners could exchange these tickets for domestic petroleum owned by other refiners, giving them a government-created source of wealth. When the import quota program was eliminated in favor of a license mechanism, this source of funds ceased, although by 1973 the value of a ticket had fallen below fifteen cents per barrel. FEA made a determined effort to ensure that the traditional import program subsidy to small refiners was retained.

FEA has concluded first that the bias initially proposed was insufficient to ensure the competitive viability of small re-

---

[63] 10 C.F.R., section 211.67.

28

finers; however, FEA has also determined that the general rationale of attempting to conform incremental entitlements issuances to small refiners to benefits received historically by this class under the oil import program is valid. Accordingly, the rule provides for a bias which is equivalent to the maximum economic preference (in the form of issuance of import tickets) received by refiners under the oil import program, with an adjustment for the inflation factor.[64]

It is ironic, of course, that the import program should be the basis of a bias in favor of small refiners since today imported crude is priced higher than domestic petroleum. Lastly, if small refiners who are required to purchase entitlements under the program can show that these payments would subject them to "serious hardship," they may be excused by means of the exceptions process from making such payments.[65]

Price regulations are also designed to preserve pre-1973 competitive relationships in the chain of distribution for petroleum products. The price regulations require that persons selling petroleum products price those products uniformly to all members of the same class of purchasers. In addition, prices to each class are based on the average selling price to the class on May 15, 1973, plus permissible increases for product and nonproduct costs. Consequently, the relative differentials *among* classes under the pricing system are supposed to remain approximately the same, as explained in one FEA ruling: "a principal function of the class of purchasers concept is to preserve the price distinctions among purchasers that customarily existed under free market conditions."[66]

The provisions dealing with the maintenance of customary class relationships are supplemented by a provision in regulations concerning the use of banked costs. In order to prevent a seller from using cost banks to increase disproportionately the prices to a particular class of customers at a later time, the regulations require that if a seller increases his price to one class of customers (since there is no rule that increases must be uniform among classes) cost banks with respect to other classes of customers purchasing the same product be debited down as if that price increase had taken place with respect to all customers for that product.[67]

---

[64] 39 Fed. Reg. 42246 (1974).

[65] For example, Northland Oil and Refining Co., 3 FEA, par. 83,063 (December 31, 1975).

[66] Ruling 1975-2, 40 Fed. Reg. 10655, 10656 (1974).

[67] 10 C.F.R., section 212.83(h).

The allocation regulations dealing with reductions of product supply during a shortage are also designed to preserve competitive relationships in the petroleum distribution chain. Section 4(c)(1) of EPAA requires that, to the extent consistent with the objectives of the allocation program, allocation regulations shall be structured to supply to each independent marketer, both branded and nonbranded, and each independent or small refiner, an amount of product equal to its 1972 product purchases. In the case of an actual shortage below those levels, supplies may be cut back only on a pro rata basis over all such businesses. Subsection (c)(2) also requires that if the allocation regulations result in a "significant increase" in the market shares of specific classes of distributors, then FEA shall undertake to revise its regulations to make an equitable adjustment.

This concern is reflected in the regulations dealing with product shortages. Under the allocation system, as will be more fully described in the next section, purchasers are granted an allocation based on an adjusted base period usage. In time of shortage, suppliers may cut back sales through the use of an across-the-board "allocation fraction" which is to be equally applied to most purchasers. Thus, all participants in the distribution network should be proportionally affected by supply shortages with respect to their base period allocation, a level which, depending on the priority of a seller's customers, may be well below current sales. Additionally, regulations require that in distributing surplus above the base period use level, suppliers of wholesale purchasers may not discriminate among branded and nonbranded independents and among outlets owned by the supplier.[68]

Allocating Scarce Petroleum Products. The current FEA regulations dealing with the allocation of petroleum products to the ultimate consumer operate under a system which was implemented during the embargo. Presently, the regulations establish, for most classes of purchasers, formulas which determine the amount of product which they are entitled to purchase. Suppliers, in turn, are entitled to receive an amount of product at least equal to the entitlements of their customers. The regulations then set out a procedure by which these allocations of products are to be reduced on a pro rata basis should supplies be insufficient to meet purchasers' entitlements.

More specifically, consumers are classified either as wholesale purchaser-consumers or as end users. Wholesale purchaser-consumers are large consumers who own storage tanks and buy large quantities of products. End users are ultimate consumers who are not wholesale

---

[68] 10 C.F.R., section 211.13(g).

purchasers, and for the purposes of some regulations, such as those dealing with gasoline, may be further divided into "bulk" end users and "other" end users.

These classes of consumers are placed into four basic priority groupings: those entitled to 100 percent of their current requirement without being subject to an allocation fraction, those entitled to 100 percent of current requirements subject to an allocation fraction, those entitled to a percentage of their base period use (often 1972 or 1973, as adjusted) subject to an allocation fraction, and those such as the motorist who are not entitled to any allocation at all. The categories which are entitled to 100 percent of current requirements without reduction are the Department of Defense, agricultural production, and space heating of medical and nursing buildings, with a large number of FEA specified "essential services" entitled to 100 percent of their current use subject to the fraction. In the case of gasoline, in which the major consumer—the motorist—receives no allocation, the station operator, classified as a wholesale purchaser-reseller, is entitled to an allocation based on his base period use.[69] Initially, base period levels could be adjusted, because of unusual growth, the amount by which 1973 usage exceeded a certain level of expected normal growth. Currently, those end users not entitled to 100 percent of their current requirements may only receive an increase in their allocation through the exception process.

Resellers notify their own suppliers of any increase in their customers' allocations and the reseller's allocation is proportionately increased. Also, resellers' allocations may be increased to reflect adjustments or assignments of customers by FEA. If a supplier is faced with a product shortage, he must determine a pro rata allocation fraction based on the allocation entitlements of those of his customers subject to an allocation fraction and the amount of product remaining for those customers after supplying customers not subject to an allocation fraction.[70] The supplier then reduces his deliveries to the customer to an amount equal to the allocation fraction times his base period volume. If a supplier's allocation fraction is greater than one, the supplier has a surplus which he can dispose of subject to certain priorities, and subject to the requirement that larger resellers must report surplus product to FEA for potential reallocation by FEA to areas of shortage.[71]

---

[69] For FEA procedures for assigning a base period level, see 1 FEA Guidelines, pars. 13,245, 13,246, appendix.

[70] 10 C.F.R., section 211.10.

[71] 10 C.F.R., section 211.10(g)(5).

State set-asides are also established under which a portion of a seller's supplies may be allocated by state energy offices to meet actual product shortages arising in those states.[72] Currently, FEA has established a Special Rule No. 3 which continues the state set-aside program for middle distillate to ensure that no marketer will lose supplies following decontrol. The regulations also give FEA the power to redirect products to portions of the country experiencing a "disproportionate" shortage. Other reserve authority permits FEA to control refinery yields to the extent technically possible, so as to mandate that refiners produce products which are in exceptionally short supply.[73]

Additional regulations provide for special cases. Separate sets of regulations are provided for the distribution of aviation fuels,[74] a procedure which must take into account peculiar needs of international air carriers, and of fuel supplies to electric utilities.[75] Unleaded gasoline is specially allocated, a procedure perhaps made more necessary by restrictive constraints on the price of that product.[76] An FEA program also allocates Canadian crude to refineries in the northern United States which are dependent on that source of supply. Canada has announced that it will phase out crude exports to the United States during the early 1980s and the regulatory scheme is necessary to provide for an orderly transition as the Canadian crude supplies are phased out.[77]

The regulations governing the allocation of propane must take into consideration that fuel's relationship to natural gas.[78] First, about 70 percent of propane is extracted from natural gas, the price of which is controlled by the Federal Power Commission. Such propane has a much lower price than propane derived from petroleum, and the consequences of these two price structures must be dealt with. Additionally, propane may serve as a substitute for natural gas. When natural gas supplies are curtailed, users may turn to these fuels as a replacement, creating shortages in other sectors. In particular, the regulations control the volume of propane inventories that businesses may accumulate and restrict the use of propane as a feedstock for synthetic natural gas plants and for peak-shaving by natural gas

---

[72] 10 C.F.R., section 211.17.

[73] 10 C.F.R., section 211.71.

[74] 10 C.F.R., sections 211.141-147.

[75] 10 C.F.R., section 211.166(d).

[76] 10 C.F.R., section 211.108.

[77] 10 C.F.R., part 214.

[78] 10 C.F.R., sections 211.81-87.

utilities.[79] Additionally, to avoid disruption which may result if supplies of imported propane are cut off, restrictions are placed on increased imports of propane to discourage import dependency. Thus, utilities must file an exception application in order to increase their imports of propane.[80]

Promoting Increased Production Capacity. As noted above, one of the principal reasons for the establishment of a multitiered pricing system for crude oil was to ensure that adequate price incentives would exist to encourage the exploration and production of new oil, while preventing windfall returns to the owners of old crude. Additional incentives for increased production of crude are provided by law and regulation. Section 8(d) of the EPAA as amended by ECPA permits the FEA to increase the average weighted price ceiling on domestic crude by up to 10 percent per annum upon a finding that such an increase would provide a "positive incentive" for exploration in high-cost areas, enhanced recovery from existing wells, and continued operation of marginally profitable wells. Also, pursuant to EPAA amended section 8(g)(2), FEA was required to submit a report to Congress by April 15, 1977, and may propose the exemption of up to 2 million barrels a day of high-cost oil shipped through the Alaska pipeline from inclusion in the national average crude price ceiling, if such exclusion is necessary to avoid compensating price decreases for the other oil included in the average, and if such reduction would limit the production of crude elsewhere in the United States. Recent amendments to the EPAA contained in ECPA exclude the first sale of oil from stripper-well properties from all price controls.

Initially, the price regulations set 1972 as the base year and production in excess of 1972 levels was classified as new crude. In order to provide an incentive for the production of increased volumes from existing wells, producers were permitted, with respect to production from a particular property, to "release" one barrel of crude from old oil price restrictions for each barrel of new crude produced from that property. Over time, this incentive feature became of decreasing value as production declined naturally from 1972 levels and granted greater incentives to workovers of existing properties than to development of new properties. Hence, in February 1976, FEA revised the crude price regulations to establish average monthly production

---

[79] See 10 C.F.R., section 211.86(g); 1 FEA Guidelines, par. 13,610; and 10 C.F.R., section 211.83(c)(2)(v). See the Glossary for a definition of *peak-shaving*.
[80] 10 C.F.R., section 211.12(g).

of old oil in 1975 as the "base period control level" (BPCL)[81] and abolished the released oil category, noting:

> many producers, especially those operating properties where current production levels are substantially below 1972 levels and further declining, were unaffected by the incentives previously afforded because those incentives became too remote to outweigh the substantial costs associated with implementing the secondary or tertiary recovery programs that might be necessary to bring current production levels up to and above 1972 levels. Moreover, the use of a BPCL based upon 1972 production levels, while well suited for use during the temporary program envisioned by the price regulations adopted initially pursuant to the Economic Stabilization Act of 1970 by the Cost of Living Council in August of 1973 and continued by FEA, is not well suited for use over the next 40 months, as it does not take into account the natural decline rate associated with primary production.[82]

In April 1976, FEA established regulations providing for semiannual adjustments to a property's base period level based on the average rate of decline in production between 1972 and 1975.[83] Under the rule, every six months, for the duration of the program, the BPCL is reduced by this constant percentage, and the amount of crude considered lower tier oil is thus decreased.

Additionally, section 122 of the recent ECPA amended section 8 of the EPAA (added by the EPCA) to require FEA to provide additional price incentives for "bona fide tertiary enhanced recovery techniques" and to adjust for arbitrary gravity/price differentials that result from current regulations.

In the event that even under these regulations a producer finds that the total revenues permitted to be received from a property under the program do not cover his costs of production, FEA will look favorably on the granting of an exception to alleviate this situation. Generally, the FEA Office of Exceptions and Appeals determines the exact ratio of upper tier to total production that would be necessary to recover costs and then permits the producer to charge upper tier prices for that volume of crude.[84]

Importantly, in 1976 the Office of Exceptions and Appeals was granting exceptions relief from the charging of lower tier prices to permit producers to earn a 15 percent rate of return on investment

---

[81] 10 C.F.R., section 212.72.

[82] 41 Fed. Reg. 4932-33 (1976).

[83] 10 C.F.R., section 212.76.

[84] For example, James M. Cunningham, Inc., 3 FEA, par. 83,176 (April 30, 1976).

in order to encourage new investment in a property. Thus, it granted an exception sufficient to generate a 15 percent return on capital needed to replace an existing well which had become unusable.[85] It has also granted an exception to permit a producer to earn that rate of return on funds invested to expand the number of wells on a property.[86]

The FEA's attitude toward increased refinery construction is clouded. On the one hand, the small refiner bias subsidizes the construction of certain refineries. On the other hand, FEA tries to prevent the buy/sell program from "subsidizing" new refinery construction. Thus, in the buy/sell program for allocation of crude, refinery capacity which was "substantially under construction" at the time of the institution of the program is eligible for automatic inclusion in the program because it is presumed that construction commitments were made without knowledge that the buy/sell mechanism would be established. With respect to "future refining capacity," inclusion in the program is discretionary with FEA and can only occur if several strict guidelines for avoiding "subsidization" are met.[87]

Likewise, under the entitlements program, new refinery capacity participates in the program when it begins operation, and existing refiners are credited with any entitlements payments accruing or owing from crude processed for them by other refiners.[88] However, FEA has opposed granting entitlement payments for crude processed by other refiners for a person who proposed to construct a refinery. Section 123 of the ECPA, added in response to this FEA attitude, requires the FEA to place a high priority on reducing regulatory barriers to entry by small and independent refiners.

The Mandatory Oil Import Program also is intended to encourage the location of refineries in the United States by charging an extra forty-two cents for petroleum products over the twenty-one cents-per-barrel fee on imported crude. FEA, however, has proposed eliminating the product fee for the duration of the crude control program because the average price for controlled domestic crude is lower than the price of imported crude by $1.80 to $3.00 per barrel. Thus, since crude costs of all domestic refiners are equalized under the entitlements program, there already is a strong incentive to utilize domestic capacity.[89]

---

[85] A & N Producing Services, Inc., 3 FEA, par. 83,172 (April 26, 1976).

[86] Austral Oil Co., Inc., 4 FEA, par. 83,004 (July 15, 1976).

[87] See 1 FEA Guidelines, section 211.65(b), appendix; Ruling 1976-1, 41 Fed. Reg. 21177 (1976).

[88] 10 C.F.R., section 211.67.

[89] 41 Fed. Reg. 30059 (1976).

Promoting Fuel Conservation and Efficient Operation of the Petroleum Industry. In general, the price and allocation regulations do not appear to have been envisioned by Congress as a method of promoting fuel conservation. Rather, this task was left to later legislation. Thus, title II of EPCA establishes standby energy conservation contingency plans and plans for rationing motor gasoline and diesel fuel to end users of those products. Title III establishes various programs for promoting energy efficiency through establishment of energy consumption standards for consumer products, revised auto standards, and targets for industrial efficiency. Indeed, the House report on the bill that was to become EPCA raised objections to the idea that price increases would reduce energy consumption:

> [T]here is even considerable doubt whether dramatically higher prices for energy will achieve their intended conservation effect. Our economy has been based upon the continued availability of cheap energy resources. We have been profligate in their use. . . . There appears to be significant room for conservation savings. Yet there is some doubt whether truly meaningful conservation or demand reduction will result from price increases.[90]

Nevertheless, various provisions of the regulations attempt to encourage full conservation and efficiency in the petroleum industry as well. As noted above, price regulations were revised to permit refiners to reallocate unrecovered costs from other products to gasoline. In part, this change was made because of the possibility that its effect might be to reduce gasoline consumption. Perhaps one of the more unusual incentives for "conservation" involves the definition of *truck* for the purposes of the gasoline allocation rules in the *Code of Federal Regulations*. Under section 211.103, businesses hauling freight by "truck" receive 100 percent of their current requirements. But, under section 211.102, a "truck" is a truck weighing more than 20,000 pounds. Thus, firms hauling freight with vehicles of less than that weight are only entitled to 100 percent of their base period allocation. One of the reasons used to justify the rule was conservation:

> The definition of the term truck and the allocation level provided for those trucks in excess of 20,000 pounds was implemented to encourage the long haul capabilities of the trucking industry, and thereby promote conservation of fuel

---

[90] H.R. Rep. No. 94-340, 94th Congress, 1st session (1975), p. 5.

36

through the use of larger vehicles which can transport materials with great efficiency than smaller ones.[91]

When United Parcel Service required an exception from the 20,000-pounds rule on the grounds that its local delivery service could most efficiently operate with smaller trucks, its application was denied since: "To the extent that UPS' operations fail to satisfy the regulatory standard, they do not promote the goals of increased fuel conservation."[92]

FEA has also attempted to influence petroleum industry output based on the relative efficiency of various uses of unfinished petroleum products in providing energy to ultimate users. In particular, FEA has determined that it should discourage the use of petroleum products as synthetic natural gas (SNG) feedstocks even though such use would help alleviate the natural gas shortage. This policy was established because energy is lost in the transformation of petroleum to SNG and the loss of energy is not in accord with the most appropriate use of the nation's energy resources:

> The manufacture of SNG from petroleum is, in most instances, an inefficient use of resources. Generally, SNG plants are designed to reform light hydrocarbons into a compatible natural gas substitute by means of a process which consumes at least 8–10 percent of the BTU's contained in the feedstock. No net increase in our supply of energy is realized. Moreover, the reformation process results in an increase in the total economic cost of available energy.
>
> . . . Accordingly, FEA will implement a policy which, in general, discourages allocation of scarce petroleum resources to manufacturers of SNG.[93]

Finally, FEA has attempted to encourage the efficient use of labor and fuel by refiners. Under current rules, a refiner can only recover increased labor costs on the basis of the number of employees on his payroll in May 1973.[94] The purpose was to discourage refineries from utilizing an increasing number of workers to process the same volume of petroleum. FEA has, however, proposed to modify this rule, stating:

> This approach [use of the May 1973 number of employees] was adopted to encourage labor efficiency—on the theory

---

[91] United Parcel Service, 2 FEA, par. 83,842 (July 30, 1975), affirmed, 3 FEA, par. 80,578 (February 13, 1976).

[92] 2 FEA, par. 83,842.

[93] 1 FEA Guidelines, par. 13,631.

[94] 10 C.F.R., section 212.83.

that new employees are associated only with increased volumes of product produced by the refinery and, thus, that the refiner would recoup the associated payroll increases through increased sales volumes. However, it now appears that more complex refinery facilities, with desulfurization equipment and complicated pollution control equipment, require refiners now to have more employees than they had for an equivalent level of output in May 1973.

FEA therefore proposes to eliminate the "incentive" aspect of this computation and to require increased labor costs to be computed in the same manner as the proposal for other nonproduct cost increases.[95]

Likewise, the rules on the passthrough of refinery fuel permitted refiners to pass on fuel costs based on the quantity of B.t.u.s utilized to refine a unit of product in May 1973. If refiners became more fuel efficient, they would get a bonus, since they could pass on costs as if they had been utilizing fuel at the higher 1973 level. Conversely, there would be a penalty if B.t.u. per unit increased. A revision to this rule is also under consideration since among other things, the increasing use of desulfurization units tends to increase per unit fuel inputs. Indeed, the rule may discourage the installation of more complex processing equipment. The rule change is being considered in the same proceeding as the labor cost limitation.[96]

---

[95] 41 Fed. Reg. 31866 (1976).
[96] Ibid.

2

THE COSTS OF REGULATION

In this chapter, the report will analyze the costs that FEA regulations in their current form impose on society: on the businessman, on the taxpayer, and on the consumer.

The Burden on Industry

The need to comply with FEA regulations imposes two types of burdens on businessmen (in addition to lost profits which may directly result from compliance with FEA price ceilings): the administrative costs of record keeping, reporting, and assisting in FEA audits, which members of the industry must incur as a matter of course; and the expense and resulting lack of flexibility caused by the need to seek FEA approval for several types of routine business transactions.

Industry Costs of Compliance. The purpose of this segment of the report is to estimate the direct industry costs of compliance with FEA regulatory programs. First, the report will discuss the likely costs of the burden placed on the industry by reporting requirements. Second, the report will estimate the total administrative costs of compliance based on estimates received from the industry. Both the reporting and the overall cost estimates are based on analytical work undertaken by the task force staff.

Reporting costs. The FEA compliance program has enforcement and audit responsibilities for the price and allocation regulations, which apply to all sectors of the petroleum industry in varying degrees. The compliance effort encompasses nine major program areas and affects more than 300,000 firms and respondents, as indicated in Table 1.

Table 1

FIRMS/RESPONDENTS SUBJECT TO FEA REPORTING AND AUDIT REQUIREMENTS, FY 1977

Program Area	Number of Firms/Respondents
Importers (crude and product) other than refiners	598
Gathering systems between crude producers and refiners	200
Independent crude producers	15,000
Major refiners (also crude producers and NGL processors)	30
Small refiners (also some crude producers and NGL processors)	110
Natural gas liquids (propane/butane) (709 plants)	123
Wholesalers (resellers) other than propane	26,000
Retailers other than propane	268,000
Retailers of propane	8,000
Total	318,061

Note: Does not include other groups (for example, municipalities) subject to reporting requirements.
Source: FEA, *Budget Amendment, Fiscal Year 1977*, p. k-2.

A key element in the success of the compliance program is the manner in which FEA monitors the industry, which is accomplished in two ways: a reporting system and an audit system for data validation.

The reporting system established by FEA and its predecessor organizations serves a variety of purposes including implementation of allocation and pricing regulations, energy development, and policy analysis. Most of the reporting is necessary to assure regulatory compliance, and the requirements are established by the regulations under authority granted by legislation. The remainder of the reporting is necessary in order to comply with the reporting requirements mandated by Congress in energy legislation. The data from these reports are used primarily for the evaluation and development of energy policy, although some serve a dual purpose, that is, both policy analysis and regulatory compliance.

A total of forty-three separate types of reports have been identified as being required, either directly to comply with the regulations

Table 2
FEA REPORTS REQUIRED FROM INDUSTRY

Type of Report	Estimated Potential Number of Respondents	Number of Reports Required			Annual Report Filings		
		Regulatory compliance	Reporting requirements of Congress	Total	Regulatory compliance	Reporting requirements of Congress	Total
One-time	Not available	4	6	10	0	0	0
As required	167,244	7	0	7	200,976	0	200,976
Semiannual	1,700	0	1	1	0	3,400	3,400
Quarterly	8,296	3	1	4	33,104	80	33,184
Monthly	36,148	11	10	21	515,004	48,372	563,376
Total	213,388	25	18	43	548,108 [a]	51,852	599,960 [a]

[a] Does not include "as required" report filings.
Source: Compiled by task force from FEA data.

41

(twenty-five) or to fulfill reporting and analysis requirements mandated by Congress (eighteen), although some may serve a dual purpose (Table 2). These reports are concerned with virtually every aspect of petroleum production, refining, and distribution, and they pertain to crude, natural gas, natural gas liquid, unfinished oils, and all products.

Only ten reports were required on a one-time basis, while another seven types of reports are filed "as required" to comply with the regulations or legislative reporting requirements. The one-time reports were submitted soon after FEA was formed to establish baseline data for providing various periodic reports (for example, market shares) required by Congress and for regulatory compliance. Although the number of these report filings is not known, it can be conservatively assumed that each of 318,061 firms/respondents filed at least one report. The cost to industry of these one-time reports can also be conservatively estimated at approximately $42 million and probably considerably more.[1]

Seven types of reports are filed "as required" on an irregular schedule for regulatory compliance. An example is the report requesting an adjustment of the base period supply volume. The potential number of respondents has been estimated at 167,244. If the average time to complete a report is 1.2 man-hours, the total annual cost, if all respondents filed only one report each, would be about $3 million (assuming a cost of $15 per man-hour).

The most important and costly category of reporting are the periodic reports (filed either semiannually, quarterly, or monthly) and submitted to either the national FEA office, regional, or state offices. These reports are filed by an estimated 46,144 respondents, which submit about 600,000 reports annually (Table 2). This reporting involves about 5.3 million man-hours at a conservatively estimated annual cost burden to industry of $79.5 million. However, almost 70 percent of this annual cost burden ($55.2 million) is the direct result of reporting requirements mandated by Congress, while the remainder is attributable to reporting required for regulatory compliance (Table 3). These data, however, are not the complete picture because they do not indicate the "as required" reports filed annually, nor do they take into account the energy reporting requirements of other federal agencies. If these latter factors are taken into consideration, the annual reporting burden to industry would, under conservative

---

[1] This figure is calculated as follows: 318,061 respondents (from Table 1) times 8.83 man-hours per report (average man-hours for annual report filings from Tables 2 and 3) times $15 per man-hour equals $42.1 million.

Table 3

ESTIMATED ANNUAL MAN-HOURS AND COSTS OF REQUIRED PERIODIC REPORTS

Type of Report	Estimated Man-Hours			Estimated Costs		
	Regulatory compliance	Reporting requirements of Congress	Total	Regulatory compliance	Reporting requirements of Congress	Total
Semiannual	0	850,000	850,000	0	$12,750,000	$12,750,000
Quarterly	22,528	800	23,328	$ 337,920	12,000	349,920
Monthly	1,598,640	2,827,368	4,426,008	23,979,600	42,410,520	66,390,120
Total	1,621,168	3,678,168	5,299,336	$24,317,520	$55,172,520	$79,490,040

a Estimated at $15 per man-hour. GAO and OMB have projected a cost factor of $15 per hour, which is conservative according to government and industry officials.

Source: Compiled by task force from FEA data.

Table 4

ESTIMATED INDUSTRY TIME AND COST TO
COMPLY WITH ENERGY REPORTING REQUIREMENTS

Agency	Number of Forms	Number of Annual Responses	Estimated Man-Hours	Projected Cost[a]
Department of Interior				
Bureau of Mines	40	52,889	80,508	$ 1,199,232
Geological Survey	25	506,734	510,475	8,032,125
FEA	50	1,887,481	8,117,769	119,066,085
Federal Power Commission	52	120,310	1,930,023	28,950,345
Total	167	2,567,414	10,638,775	$157,247,787

[a] The cost estimate was based on a man-hour cost of $15.
Source: Commission on Federal Paperwork, Position Paper, September 24, 1976.

assumptions, approximate $157 million, and FEA reporting would alone cost $119 million, or about 76 percent of the total (Table 4).

The four federal energy agencies (that is, Bureau of Mines, Geological Survey, Federal Power Commission, and FEA) combined utilized about 167 forms soliciting over 2.5 millon annual responses, which take an estimated 10.6 million man-hours to complete. Much of this reporting has been criticized for the "lack of credible data, overlapping and duplicative reporting requirements, and confidentiality implications," which indicates a strong need for improvement.[2] The General Accounting Office (GAO), Office of Management and Budget (OMB), Commission on Federal Paperwork, and the Federal Inter-Agency Council on Energy Information have recognized these criticisms, and some progress, however slow, has been made. For example, in July 1976, the FEA and the Bureau of Mines finalized an agreement for a joint report to obtain monthly data on refineries, bulk terminal product stocks, product pipelines, and crude oil stocks. This joint report is expected to effect some major cost savings for industry. Efforts to alleviate overlapping and duplicative reporting, as well as to resolve the confidentiality issue, are expected to continue in the future.

If all energy reporting since the Arab oil embargo is taken into consideration, the total reporting burden to industry through the end

[2] Commission on Federal Paperwork, Position Paper, September 24, 1976.

of fiscal year (FY) 1976 would approximate $357 million. This estimate was derived as follows:

Annual reports FY 1975 and FY 1976 (all agencies)	$314.5 million
One-time FEA reports	42.1 million
Total	$356.6 million

The reporting burden imposed on industry, however, represents only a part of the total cost burden to industry of complying with energy regulations. Other costs are discussed in the next section.

Most of the current annual reporting required for regulatory compliance focuses on the allocation regulations. As allocation and pricing controls are lifted on various petroleum products, FEA's data-gathering activities will be reduced accordingly. For instance, in July 1976, when middle distillates were decontrolled, FEA reviewed its reporting requirements in this area. A full-scale reporting system requiring monthly reports from approximately 4,000 firms selling number 2 heating oil was discontinued. Because of congressional concerns about possible large price increases of number 2 heating oil, however, FEA designed a new data system to monitor heating oil prices. Since the needs of this type of study vary considerably from those required for a compliance type study, FEA was able to keep the reporting burden at a minimum. FEA, in fact, by employing statistically valid sampling techniques, was able in this case to limit to 600 the number of firms required to supply the necessary data. In contrast to the 4,000 or so reports filed monthly under the old "control" system, these new requirements represent a significant reduction in the respondent's reporting burden. In any case, as FEA shifts its emphasis to statistical and general monitoring types of studies, sampling techniques and other revised and improved survey procedures will, it is anticipated, significantly decrease the overall reporting burden, especially on small businesses. Presurvey investigation, including the pretesting of proposed survey forms with potential respondents, will further enhance FEA's ability to address the problem of the reporting burden.[3]

---

[3] According to FEA, the pretesting procedure basically involves requesting a *small* sample (ten to fifteen) of potential respondents to a proposed survey to provide voluntarily comments about the proposed reporting vehicle, that is, FEA does not normally ask for any actual data, just whether the firm can provide the data, what its reporting burden will be, suggestions for changes, improvements, et cetera.

Overall administrative costs. Substantial administrative costs are incurred by the petroleum industry on an ongoing basis in order that FEA's personnel can carry out their compliance and enforcement responsibilities.

Administrative costs to firms in the petroleum industry include, but are not limited to, such costs as those incurred due to: (1) diversion of personnel (to tasks essential to the firms' "effort toward compliance") as well as the hiring of additional personnel; (2) additional record keeping resulting from FEA price and allocation regulations; (3) FEA reporting requirements. In addition, many firms incur substantial legal costs as a direct result of the regulations; for example, legal fees for compliance advice, applications for exceptions, litigation. Administrative costs are direct costs and by definition would not include any opportunity costs which society might have incurred because of FEA's price and allocation regulations.

Efforts to identify and quantify administrative costs to various segments of the industry are difficult because these expenditures seem to permeate each and every phase of the industry's operations. Although difficult, this cost information is not impossible to obtain. It is interesting to note, however, that virtually no information was available within the agency as to the administration costs incurred by firms in attempting to comply with the price and allocation regulations. Such information should be gathered by the agency in order that it be able to assess fully the impact of its regulation on those who are regulated.

A comprehensive study to determine the administrative "cost of regulation" has not been performed. However, certain examples of the magnitude of these costs, at the various levels of the industry, were obtained by the following means: contacts were made with several firms in the petroleum industry, both by personal visit and by telephone; contacts were also made with FEA advisory committees, trade associations (for example, American Petroleum Refiners Association), the GAO, and certain members of Congress.

The results of these contacts (that is, information obtained from those choosing to respond to the task force's request for cost data) are summarized below. Cost information obtained is presented for each segment of the industry. These estimates should be compared with the estimates of reporting costs, developed by the task force staff on a more detailed basis, which were presented in the preceding section.

Producers: One producer, responding to a task force request for an estimate of administrative costs, estimated that the firm would expend $165,000 during calendar year 1976 in an effort to comply with FEA regulations. According to the producer, these costs would relate to: (1) full- and part-time personnel; (2) FEA report preparation; (3) legal services; (4) computer programming; and (5) consulting services. This information does not provide a sufficient basis for projecting the overall costs for this segment of the industry.

Refiners: Estimates of the administrative cost of compliance were obtained from ten U.S. refiners. These firms estimated their annual expenditures to be in excess of $52 million.[4] Based on these cost estimates, industry-wide projections of the "cost of compliance" can be made.

U.S. refiners have refining capacity totaling 18 million barrels of crude oil per day or 6.6 billion barrels per year.[5] Refiner responses as to the cost of regulations indicate administrative costs might range from 0.9 cents to 8.6 cents per barrel of refining capacity. Based on these estimates, it appears the total cost of regulation to the petroleum industry's refiners, assuming a 100 percent effort toward compliance, could be as high as $570 million annually.

It should be noted, however, that with regard to the larger firms, these costs may not represent additional costs because many refiners maintain a permanent capacity to perform these types of compliance activities for the Securities and Exchange Commission, state regulatory agencies, et cetera.[6] Thus, it is conceivable that this capability would be maintained to some extent, with or without the FEA price and allocation regulations.

Marketers: Only one response was received as the result of our contacts with FEA's marketing advisory committees. A trade association for marketers of fuel oil in New England estimated the cost of compliance to range from 0.5 cents to 1.1 cents per gallon of product sold. This data was felt to be representative of administrative costs incurred by this segment of the industry as it is a trade association estimate and not that of an individual firm. As such, it may be satisfactory for use as a base for estimating the cost of FEA regulations on marketers.

---

[4] Aggregate administrative cost estimates for Apco Oil Corporation, Atlantic Richfield Company, Continental Oil Company, Exxon Company U.S.A., Marathon Oil Company, Phillips Petroleum Company, Shell Oil Company, Skelly Oil Company, Standard Oil Company of California, and Standard Oil Company of Indiana.

[5] FEA, Office of Regulatory Programs, *Capacity List, The Refiners and Refineries of the United States*, 6th ed., 1976.

[6] Interview with Charles Owens, formerly deputy assistant administrator of FEO's Office of Policy, Planning and Regulation, September 29, 1976.

Marketers (for example, resellers, jobbers, retailers) compose the largest segment of the petroleum industry. The total number of these firms is about 300,000, and they range from large, computerized, high-volume operations to the small "mom and pop" business. Each year these firms sell a large share of refined petroleum products. For example, in calendar year 1975, independent marketers sold 54.9 percent of all the distillates sold in the United States. (The remainder is sold by refiner/marketers).[7]

Based on the estimate of 0.5 cents to 1.1 cents per gallon for cost of compliance, and assuming a 100 percent compliance effort (that is, all firms make some effort toward compliance with the price and allocation regulations), the cost of regulations for fuel oil marketers could be estimated to range from $195 million to $428 million annually.[8]

Conversely, assuming only 2 percent of the firms currently make any effort toward compliance (that is, to keep adequate records, including computations of maximum lawful selling prices), administrative costs might be negligible: $4 million to $9 million per year.[9]

Administrative costs for gasoline marketers could also be estimated in the above manner. One estimate is that the administrative costs to all firms in the petroleum industry, large and small, is approximately 1 percent of annual revenues.[10] Moreover, it appears the greatest burden of FEA regulations is on small marketing firms because of their limited "compliance capability," that is, the capability which these firms can feasibly and economically maintain.

Interference with Normal Distribution Patterns, and Creation of Competitive Inequity and Uncertainty. The FEA regulatory structure not only imposes direct costs on businessmen but also interferes with their freedom to respond to the forces of the marketplace. First, regulations force businessmen to seek government approval of ordinary business transactions which, in the pre-FEA era, were of little concern to other than the interested parties. Because of the supplier/purchaser freeze, agency approval must be obtained before specific changes in business relationships can be made. Also, because of the regulatory requirement to obtain FEA approval of a new relationship, businessmen

---

[7] Discussion with Jim Diehl, Bureau of Mines, Office of Fuels Data.

[8] These figures are calculated as follows: 38.9 billion gallons of distillate and residual fuel sales (excluding sales by refiner/marketers) times 0.5 cents per gallon equals $195 million; times 1.1 cents per gallon equals $428 million. Bureau of Mines, Office of Fuels Data, *Annual Survey of Fuel Oils and Kerosene*, 1975.

[9] Interview with Gordon Harvey, director of Compliance Policy and Planning, FEA Office of Regulatory Programs.

[10] Interview with Charles Owens.

must request the assignment of suppliers and base period volumes, even though "surplus" supplies are readily available.

Second, the operation of FEA regulations causes competitive inequalities among new and old businesses and inhibits the forces of the marketplace from operating. Unfortunately, the administration of these regulations in a time of surplus has undermined the rules' ability to work in a time of shortage.

FEA regulations require government intervention in the normal conduct of business. FEA regulations provide that the relationship between a supplier and a wholesale purchaser-reseller may not be waived or terminated without the written approval of FEA, and that the relationship between a supplier and a wholesale purchaser-consumer may be terminated by mutual consent.[11] Likewise, FEA must be notified of the establishment of a relationship between a new supplier and a new wholesale purchaser-consumer.[12] The new relationship is subject to FEA modification. Establishment of a base period relationship for a new wholesale purchaser-reseller requires FEA approval. In addition, if a wholesale purchaser is unable to find a willing supplier, it may apply to FEA for an assignment of a base period supplier.

Some idea of the burden which the regulations place on businessmen may be had by looking at the number of routine transactions pursuant to the regulations handled by regional operations offices. It is estimated that over the life of the program 300,000 to 400,000 applications have been filed with FEA regional offices. For example, a purchaser requesting an adjustment in base period volume or assignment of a supplier and base period volume must file a form FEO-17 with the regional office, or with the national office in the case of air carriers and certain other parties. In the fall of 1975 and the spring of 1976, for example, regional offices were receiving, on the average, two thousand to three thousand new cases a month, of which five hundred to eight hundred resulted in new assignments of base period suppliers and volumes and two hundred to three hundred per month in approval of mutual terminations of relationships.[13]

In a December 4, 1974, teletype message, the head of the Office of Exceptions and Appeals gave directions to regional offices about when regional offices could approve change-of-supplier transactions as part of a routine (operations) procedure, and when the applicants

---

[11] 10 C.F.R., section 211.9(a)(2).

[12] Ibid., section 211.12(e).

[13] Compiled by task force staff from FEA data.

had to file for an exception. The 1974 directive permits the termination of a supplier-purchaser relationship for a wholesale purchaser-reseller and the assignment of a new relationship if there is a three-party mutual agreement between the old supplier, the new supplier, and the purchaser. In the case of wholesale purchaser-consumers, mutual consent automatically terminates the supply obligation, but a new supplier is assigned only if that supplier agrees in writing to assume supply obligation.

Other than in the situation specified above, an exception is required to terminate a base period relationship and reassign the wholesale purchaser involved to a new supplier.[14]

Perhaps one of the most complex interactions between FEA and businessmen involves the sale of a refinery. Refiners have many purchase and supply obligations, entitlements obligations, and obligations to other refiners under the buy/sell program. Changes in these obligations resulting from a sale are made all the more difficult because the obligations under many programs are obligations of the company owning the refinery, and they do not "run with" the refinery when it is sold. Additionally, since entitlements under the small refiner bias and status under the buy/sell program do not change on a strictly proportional basis when a refiner's capacity changes, complicated adjustments must be made in obligations under these programs. FEA has no formal guidelines for dealing with the sale of a refinery and the procedure is handled under the exceptions process.

Even when a refinery shuts down its operations for a month for extended maintenance work, many of its pricing and supply obligations will be affected. It must, therefore, seek exception relief from FEA before beginning maintenance work in order to clarify the impact of the shutdown on its regulatory obligations.[15]

FEA regulations distort competitive relationships. Perhaps the clearest example of the needless regulatory interference in business practices which the regulations require is the assignment of a base period volume and supplier for "new" retail outlets. In July 1976, there was a backlog of approximately twenty-four hundred assignment requests in the regional offices, most of which were for the assignment of base period relationships for motor gasoline. An August 12, 1976, survey of regional offices undertaken for the national FEA Office of General Counsel indicated that retail outlet assignment cases by all

---

[14] Teletype from Melvin Goldstein to FEA regional administrators, December 4, 1974 (national office file copy).

[15] For example, Union Oil Co., 3 FEA, par. 83,105 (February 20, 1976).

regions had exceeded one thousand cases per month during the four months ending in July 1976.

In May 1975, an elaborate set of guidelines was published for evaluating applications for new gasoline retail sales outlets.[16] These guidelines theoretically provided for notice to and comments from competing retailers and other parties who might be adversely affected by the new assignment. According to operations personnel in some regional offices, these "due process" requirements have substantially increased the administrative complexity of handling such cases.

The regions estimated that the average cost per case to implement the prescribed newspaper notices to aggrieved parties was approximately $260. At the present level, then, of twelve thousand cases annually, the newspaper notices required by FEA guidelines would cost more than $3 million per year. However, because of the lack of funds in the regions, the prescribed procedures are not being followed.

During the first few months of the regulatory program when procedural instructions were minimal, most assignments of base period volumes and supplies were handled on an assembly line, rubber stamp basis. Although current procedures are formally much more in accord with proper administrative procedure, the example of this $3 million-per-year media notice requirement lends strong support to the proposition that the procedures remain unnecessary and wasteful. Indeed, FEA appears de facto to have obviated the rationale for requiring approval, since the guidelines begin with the strong presumption that assignments to new retail outlets should be granted. FEA believes that if market conditions warrant the opening of a new station, FEA personnel should not obstruct its opening and thereby deny a man his "right" to go into business. On July 9, 1976, the FEA national office did send a message to the regional offices cautioning against overly liberal assignments which distort the allocation mechanism.

In essence, the new-station assignment procedure has become primarily a burden to the businessman attempting to open a new station, a workload creator for regional office personnel, and a would-be subsidy to local newspapers. What the procedure is not, however, is an effective program for establishing appropriate patterns of gasoline distribution during a shortage period.

Rather than serving to rationalize gasoline distribution in a shortage, the program's major effect is to create inequities between old and new service stations. Under the allocation guidelines, new service stations are granted a base period volume equivalent to the

---

[16] 1 FEA Guidelines, par. 13,245, appendix.

allocation entitlements usage of similar stations in the market area, while existing stations are limited to their 1972 base period volumes (as possibly adjusted by the "unusual growth" and "changed circumstances" rules of 1973 and 1974). Consequently, there is concern in the business community that, should a shortage occur, established businesses will be placed at a relative disadvantage as to product availability in comparison with newly opened stations. This concern is particularly present among service stations which have undergone a change in their marketing methods that has resulted in a sharply increased sale of gasoline. Were a gasoline station operator to close down his present relatively low-volume, full-service station, and to open up a high-volume "gas and go" station a few blocks away, the operator would be able to get an assigned allocation entitlement on the basis of the allocation entitlement of a similar high-volume station. On the other hand, were the operator to remain at the same location and renovate his current station, he would only retain his base period entitlement.

In order to deal with this inequity, and in the absence of guidance from headquarters, regional offices undertook to develop criteria about when a renovated station constituted a "new" outlet. According to a 1975 survey taken by the FEA Office of General Counsel, all ten regions permitted a station which physically closed down during renovation to apply for a new station allocation. However, the time for which a station needed to be closed varied from region to region.[17] Thus, if a station managed to keep a single pump operating during its renovation, it would not be entitled to a revised allocation even if its method of operation changed entirely.

On August 31, 1976, FEA published Ruling 1976-5.[18] This ruling terminated the regional office practice of granting new outlet status for renovated stations. The ruling held that only if a gasoline station legitimately went out of business—that is, its owners intended to cease operating the property as a gasoline station—would a subsequent renovation of the property entitle the operator of the property to receive an allocation as a new outlet. The ruling stated that the only ways in which the owner of a renovated station could receive gasoline sufficient to operate his new type of station would be to purchase surplus product or to request an exception from the Office of Exceptions and Appeals.

---

[17] FEA, Office of Allocation Regulations, "Changes Needed in Motor Gasoline Regulations," Issue Paper, 1976.
[18] 41 Fed. Reg. 36647 (1976).

Of course, the obtaining of an exception is a much more complicated process than the submission of a Form FEO-17 to the regional operations office. Much more documentation is required and a showing of "serious hardship" or "gross inequity" is generally necessary before the exception will be granted.[19] The effect of Ruling 1976-5, consequently, will be to perpetuate distinctions between long-time service station owners and new entrants and will unfairly discriminate against the former if a shortage should occur while current regulations are in force.

The regulatory system may also work to effect a discrimination between new and old operations conducted by one business. For example, on May 6, 1975, United Parcel Service (UPS) requested an exception to increase its base period allocation in two ways. First, it requested an increase in the quantity of motor gasoline allocated to it for operation in the forty-three states which it served prior to July 1975. Second, it requested an allocation of gasoline to cover its projected needs in five additional states which it had not previously served but was authorized to serve by the Interstate Commerce Commission after July 1975.

On July 30, 1975, the Office of Exceptions and Appeals issued an order denying UPS's request for an increased allocation to cover operations in its traditional states, but granting the request for an allocation for the new states at 100 percent of UPS's estimate.[20] The decision indicated that the expected shortfall in the traditional states only amounted to about 10 percent and thus did not meet the hardship or inequity requirements, particularly in view of the availability of surplus product. On the other hand, the lack of any allocation for gasoline usage in the new states clearly merited relief. As a result, in a time of shortage, UPS might well be limited to 90 percent of its current operations in the traditional states, but could conduct business as usual (on a 1975 basis) in the new states. UPS appealed this decision on September 29, 1975, and the initial decision was upheld on February 13, 1976.[21]

The operation of the program itself, even without regard to the manner in which FEA officials will handle particular operations or exceptions applications, may cause uncertainty and loss of competitive opportunity for the businessmen. Sellers in many cases find that the customers whom they are obligated to supply do not purchase all of their allocation and, instead, purchase surplus product elsewhere.

---

[19] See 10 C.F.R., section 205.54.

[20] United Parcel Service, 2 FEA, par. 83,262 (July 30, 1975).

[21] 3 FEA, par. 80,578 (February 13, 1976).

Even though such "underlifting" of product may continue for some time, sellers must take care to see that they have sufficient product on hand to meet the request of those purchasers should they actually choose to exercise their allocation rights. On the other hand, purchasers may find themselves unable to switch suppliers in those markets where there is less surplus, since the base period supplier may like to keep a particular customer and a new supplier does not wish to take upon itself the obligation of supplying an additional customer in a shortage. As a result, many dealers are prevented from shopping around to find the best price for product. It was for this reason that an association of New England distillate and residual fuel oil dealers strongly lobbied Congress to permit FEA to remove regulations over those fuels. They reported that the rules made them captive to their present distributors and increased the general price level since suppliers recognized their customers were inhibited from seeking new suppliers.[22]

The Burden on the Taxpayer

The purpose of this section is to identify the direct administrative costs of FEA regulation to the government and, therefore, to the taxpayers. These costs are considered to be past and projected FEA budget outlays.

The FEA officially came into being on June 26, 1974, when the Internal Revenue Service (IRS) transferred control of the regional compliance force to the FEA regional administrators. IRS also transferred control of compliance reporting and case control systems to the FEA national compliance office. Prior to that time, regulatory activities were carried out by the FEO, the CLC, the IRS, and personnel temporarily assigned to FEO from other federal agencies. Because regulatory activities were widely dispersed, it is difficult to derive a precise estimate of regulatory budget outlays prior to the formal establishment of FEA. However, Frank G. Zarb, the FEA administrator, has estimated that "allocation during the recent embargo required the full-time efforts of about 4,000 people and cost approximately $100 million; in addition, substantial record keeping, reports and audits were required of the private sector."[23]

---

[22] New England Fuel Institute, "Position Paper on Middle Distillate Decontrol," presented to the members of the House and Senate from New England, June 29, 1976.

[23] U.S. Congress, House of Representatives, *Hearings on Energy Conservation and Oil Policy, before a Subcommittee of the House Committee on Interstate and Foreign Commerce*, 94th Congress, 1st session (1975), part I, p. 196.

Through the end of FY 1976, actual FEA obligations totaled $339 million and actual man-years of effort amounted to 7,121 (Table 5). Adding the $100 million previously estimated by Mr. Zarb means that obligations through the end of FY 1976 totaled approximately $439 million, which is the most direct measure of outlays by the government (and therefore the taxpayers) in meeting the requirements of regulatory legislation. For FY 1977, obligations are estimated to increase to $150 million, with the Strategic Petroleum Reserve Program requiring an additional $743 million, bringing the total to $893 million.

Man-years of effort grew from 3,139 in FY 1975 to 3,386 for FY 1976. For FY 1977, the total is estimated at 3,689 man-years, or 17.5 percent above FY 1975.

"Regulatory Programs" is the largest program element in the FEA budget. For FY 1976, for example, Regulatory Programs accounted for 28 percent of obligations (up from 24 percent in FY 1975) and 41 percent of total man-years (Table 5). Comparable figures for FY 1977 are estimated at 29 percent and 46 percent, respectively (excluding the Strategic Petroleum Reserve Program). This program element is especially important because it has the primary responsibility for executing regulation development, allocation, compliance, and contingency planning responsibilities of the FEA.

The major activities within the Office of Regulatory Programs (ORP) center around compliance and operations, which, when combined, accounted for more than 90 percent of obligations and man-years in FY 1975 (Table 6).

The compliance program encompasses the following activities:

- Developing and promulgating enforcement policies, plans, and audit guidelines to regions;
- Developing, scheduling, and supervising training for auditors, investigators, and program managers;
- Providing technical assistance and supervision in conduct of audits and investigations;
- Providing technical assistance and supervision and securing required legal support in the formal resolution of cases involving violations;
- Maintaining continuous monitoring of all compliance activities, preparing required reports, and providing required management direction to regions;
- Carrying out field audits and investigations and completing formal action on all cases. (This function is executed by regional personnel.)

Table 5

FEA SUMMARY OF OBLIGATIONS AND MAN-YEARS, FISCAL YEARS 1974–1977

	1974 Actual	1975 Actual	1976 Actual	1977[a] Estimate
	Amount ($ in thousands)			
Executive direction and administration	N.A.	46,027	35,783	31,454
Energy policy and analysis[b]	N.A.	21,769	22,463	28,446
Regulatory programs	N.A.	32,132	37,164	43,337
Energy conservation and environment	N.A.	17,020	23,847	35,163
Energy resource development	N.A.	13,767	10,793	10,404
International energy affairs	N.A.	1,291	1,162	1,581
Subtotal	N.A.	$132,006	$131,212	$150,385
Strategic petroleum reserve	N.A.	N.A.	1,818	742,781
Total obligations	$73,833	$132,006	$133,030	$893,166
	Man-Years			
Executive direction and administration	N.A.	908	1,044	930
Energy policy and analysis	N.A.	366	349	369
Regulatory programs	N.A.	1,285	1,396	1,625
Energy conservation and environment	N.A.	255	270	294
Energy resource development	N.A.	284	271	290
International energy affairs	N.A.	41	41	44
Subtotal	N.A.	3,139	3,371	3,552
Strategic petroleum reserve	N.A.	N.A.	15	137
Total man-years	596	3,139	3,386	3,689

[a] The new fiscal year for all federal agencies will begin on October 1 in fiscal years beginning with FY 1977. Estimates for the transition quarter are not included.

[b] Function divided in mid-1976 and now includes Energy Information and Analysis (EIA) and Policy and Program Evaluation (PPE).

Sources: FEA, *Budget Amendments, Fiscal Year 1977;* and FEA, Office of Budget and Financial Management.

Table 6
ORP OBLIGATIONS AND MAN-YEARS, FY 1975

Program	Actual Obligations ($ in thousands)	Actual Man-Years
Headquarters		
Compliance	$ 2,815	59
Operations	2,474	120
Contingency planning	0	0
Regulatory development	1,790	67
Administration	629	32
Data support	435	18
Total, headquarters	8,143	296
Regions		
Compliance	20,227	784
Operations	3,762	205
Total, regions	23,989	989
Total, ORP	32,132	1,285

Source: FEA, *Budget Amendment, Fiscal Year 1977.*

As indicated in the FY 1975 budget for ORP, most of the compliance activities, as well as those in operations areas, are carried out in the regions, which accounted for about 75 percent of the ORP budget in terms of both obligations and man-years (Table 6).

The operations function of ORP is currently charged with administration of the following programs:

- Crude Oil Entitlements Program
- Domestic Crude Oil Allocation Program (Buy/Sell)
- Canadian Crude Oil Allocation Program
- Refinery Yield Program
- Mandatory Oil Imports Program
- Propane/Butane and Other Products Allocation Program
- Motor Gasoline Allocation Program [24]

---

[24] The enactment of EPCA, section 454, is assumed to result in the phase-out of this program in FY 1977 due to the exemption of specified allocated petroleum products from the provisions of EPAA, section 4(a), at the wholesale and retail levels.

- Middle Distillate Allocation Program [25]
- Aviation Fuels Allocation Program [26]
- Residual Fuels Allocation Program (Utilities) [27]
- Residual Fuels Allocation Program (Non-Utilities) [28]
- Regional Assistance Program [29]

Programs targeted for decontrol in FY 1976/1977 have been phased down from end FY 1976 levels to end FY 1977 in anticipation of decontrol and in recognition of a reduced workload because of plentiful product supplies.

In addition to compliance, operations, and other activities within ORP, other direct regulatory activities are carried out by the Office of General Counsel (which provides all legal support) and the Office of Private Grievances and Redress, under which is included the Office of Exceptions and Appeals. Aggregating all these activities indicates that total obligations for direct regulatory activities amounted to $33.7 million in FY 1975, increasing to $41.0 million in FY 1976; 1977 obligations are estimated at $47.2 million, or about 40 percent above the 1975 level (Table 7).

For General Counsel, the increase in 1976 was primarily due to expanded responsibilities resulting from the enactment of EPCA (in December 1975); for Private Grievances and Redress, the increase reflects growth in the number of exceptions and appeals applications processed. A relatively small increase in the FY 1977 budget for Private Grievances and Redress is anticipated. The FY 1977 budget also anticipates that there will be a very significant reduction in the number of allocation cases which will be filed with the Office of Exceptions and Appeals.

The Economic Costs of FEA Regulations under Present Conditions

This section of the report analyzes the extent to which consumers are likely to benefit under the regulations during the current period of crude oil and refined product surpluses. The economic effects of the regulations in a period of shortage are not discussed.[30]

[25] Ibid.

[26] Ibid.

[27] Ibid.

[28] Ibid.

[29] Ibid.

[30] See the Compliance Program section of this report below for an analysis of the *ability* of the current program to work in a shortage.

Table 7

OBLIGATIONS FOR REGULATORY PROGRAMS WITHIN FEA OFFICES, FISCAL YEARS 1975–1977

($ in thousands)

Office	1975 Actual Obligations	1976 Actual Obligations	1977 Estimated Obligations
Office of general counsel	749	2,186	1,941
Office of private grievances and redress	832	1,675	1,894
Office of regulatory programs			
Compliance	23,042	24,782	43,337 [a]
Operations	6,236	9,232	—
Other	2,854	3,150	—
Total	33,713	41,025	47,172

[a] This figure is a total for all ORP obligations.

Sources: FEA, *Budget Amendments, Fiscal Year 1977;* and FEA, Office of Budget and Financial Management.

FEA regulations were designed first to prevent producers from capturing windfall gains on their crude oil supplies, and then to force refiners and marketers to pass the savings through to consumers.

A second major goal of energy policy has been to encourage capacity expansion at a rate that permits a high proportion of U.S. demand to be satisfied by domestically refined products. Expansion of domestic capacity to keep pace with expansion of domestic demand would prevent undue reliance upon foreign sources of product supplies, and would help maintain the competitive viability of the independent refining and marketing sectors.

Despite these intentions, however, domestic refining capacity is not being encouraged to expand, and the savings in crude oil costs are not being fully passed through to the final users of refined products. Capacity is not encouraged to expand because the major incentive mechanism, the entitlements program, is only a temporary phenomenon that will expire with crude oil price controls in 1979. In addition, other aspects of the regulations, particularly the cost passthrough rules, actively discourage addition of capital, whether to expand total capacity or to change the ability of a given-sized refinery to produce particular products.

With demand for refined products certain to grow as economic activity expands, the increase in output can only come from two sources: more intensive use of existing domestic capacity, or product imports. The present regulations, however, pose a significant barrier to product imports. The result will, therefore, be an increasingly intensive use of existing domestic refining capacity. This, in turn, will lead to rising prices for consumers, as increasing the use of existing capacity beyond a certain level leads to rising out-of-pocket costs for additional units of output. Moreover, with still further intensive use, these costs will rise quite rapidly. The resulting higher consumer prices are reinforced by other aspects of the present FEA regulatory structure, particularly the bias towards small refineries, the freeze in supplier-purchaser relations, and the class-of-purchaser restrictions.

The implications of these trends can already be seen in refined product prices. Spot market prices for most of the period to date have been higher than would have been the case if the entitlements program had not blocked imports of refined products.

FEA Regulations Discourage the Overall Expansion of Domestic Refining Capacity, and Encourage the Capacity Likely To Be Built To Be Less than Optimal Size. Domestic refining capacity that is ample to keep pace with growing domestic demand for refined products has two benefits for consumers. In terms of national energy policy, any embargo is more easily dealt with if there is ample domestic refining capacity than if the majority of imports are in the form of refined oil products. It is easier to try to locate alternative sources of crude oil and to make the necessary alterations in refineries to refine crude oils of differing properties than it would be to try to build new refineries to replace those closed to the nation by an embargo.

A second major benefit of having ample domestic refining capacity in relation to domestic demand is that it enhances the competitive viability of the independent sectors of the industry. The encouragement of domestic refining over foreign offers an incentive for would-be refiners who do not have guaranteed access to domestic crude oil supplies to enter the industry and import crude oil for their refineries. When capacity is ample relative to demand, both independent and integrated refineries serve as sources of supplies of refined products for independent marketers who compete with the market outlets of the integrated companies. Both the independent refiners and the marketers help to preserve the benefits of competition for consumers in the form of the lowest possible prices.

For both these reasons, therefore, it is desirable to have a domestic refining industry that expands as domestic demand grows. The present FEA regulatory structure discourages the growth of capacity.

Cost passthrough rules discourage investment in refining. The FEA cost passthrough rules on pricing refinery output forbid refiners either to pass through or to bank many of the increased costs of capital. Refiners may pass through increased interest charges, but not increases to provide a return on additional capital that is not borrowed. The proposed rulemaking to allow passthrough of depreciation allows only for recovery of the initial amount of the investment, but still no return on that capital.

The disincentive to invest created by the cost passthrough rules is made worse by the fact that the cost of expanding capacity has risen since the imposition of controls. The Nelson refinery construction cost index rose from 438.5 in 1972 to 601.8 in May 1976 (it was 100 in 1946).[31] Failure to increase the allowable margins in the price of refined outputs means that the higher capital costs of refining cannot be recovered. This, in turn, acts as a brake on capacity expansion.[32]

The disincentive toward expansion of capacity provided by the failure of the cost passthrough rules to allow a rate of return on new capital is reinforced by the fact that the freeze on supplier-purchaser relations, the allocation of new customers to a supplier, and the class-of-purchaser rules may work together to prevent a refiner from selling an additional unit of output to the customer who is willing to pay the highest price. Therefore, he may be unable to charge an additional customer a price that covers the cost of increasing his output sufficiently to supply that additional customer. Normally, when the costs of additional output are rising and are translated into prices, this serves as an incentive to expand capacity. The regulations work to prevent this incentive from working. In such a situation, additions to capacity are likely to be discouraged, and those that are made are likely to be not only smaller than optimal but also designed to produce products in a mix different from that which the market would evoke.

The biases toward small refiners in the existing regulations tend to encourage capacity expansion projects which are smaller than optimal with respect to available economies of scale. The various small refiner biases in the present regulations phase out at 175,000 barrels

---

[31] *Oil and Gas Journal*, October 4, 1976, p. 96.

[32] Gorman Smith, assistant administrator for regulatory programs, in an interview on October 13, 1976, stated that the regulations have worked to reduce refinery expansion.

per day of capacity. If the present regulations were continued, new entrants into refining would face strong incentives to restrict the size of their refineries to fall under that size limit. Without the regulations, however, market forces would tend to induce new entrants to try to build larger refineries in order to take advantage of economies of scale in refining. According to one study, the minimum size refinery necessary to take full advantage of economies of scale is one designed to process 200,000 barrels per day. Building a refinery with a capacity which is one-third of that size, or about 67,000 barrels per day, over the long run raises average unit costs by about 5 percent.[33] Nevertheless, the large subsidy that small refiners receive from the entitlements bias is a very powerful incentive to build smaller refineries.

The present entitlements program operates as follows: The price of an entitlement is defined to be the difference between the weighted average cost to refiners of "old" oil and of imported crude oil, less twenty-one cents. "An entitlement is the right to process 'deemed old oil' which is the sum of a refiner's receipts of 'old' oil and a fraction of his receipts of 'upper tier' crude oil."[34] A refiner must purchase entitlements for the amount of "deemed old oil" used in excess of the national domestic crude oil supply ratio (NDCOSR) times his crude runs to stills.[35] If a refiner uses less old oil than this ratio, he will sell entitlements. Because all refiners are entitled to the same amount of "deemed old oil" as a proportion of their total crude runs, the average cost of crude is ideally the same for all refiners.[36]

The entitlements program is not, in fact, a neutral program that merely equates the cost of crude oil for all refineries. Rather, small refiners get additional entitlements. Table 8 gives data on the entitlements program. Tables 9 and 10 show the value of the small refiner

---

[33] Frederic M. Scherer et al., *The Economics of Multi-plant Operation: An International Comparisons Study* (Cambridge: Harvard University Press, 1975), p. 80. The figures are for a refinery built in the United States.

[34] FEA, *Monthly Energy Review*, October 1976, definitions section. The fraction is set monthly by FEA. This fraction multiplied by the price of an entitlement should equal the differential between average price of imported crude and average price of upper tier crude. In other words, the right to process a barrel of upper tier oil is worth this fraction of the right to process a barrel of old oil.

[35] FEA, *Monthly Energy Review*, July 1976, p. 78.

[36] In fact, the system does not work to make all refiners face the same cost for crude oil even ignoring the small refiner bias, as imports of crude oil do not all cost the same. A refiner who can get most of his imports for less than the average price gains under the system; a refiner who pays more than the average price loses. Based on FEA proprietary data, the variation in the average cost of crude oil net of entitlements for sixteen large integrated refiners in January 1976 ranged as high as $3.26 per barrel. ICF, Inc., *Final Report: Banked Costs and Market Structure and Behavior*, June 1976, p. 11.

Table 8

MONTHLY IMPACT OF ENTITLEMENTS PROGRAM

(average of November and December 1974 data)

All refiners and importers	
Total of all refiners listed	132
Total entitlements issued	15,429,363
Total entitlements exempted by Special Rule No. 3	3,823,635
Total entitlements issued but for Special Rule No. 3	19,252,998
Value of total entitlements (issued and exempted)	$96,264,990
Small refiners (under 175,000 capacity)	
Total small refiners listed	112
Small refiner buyers	20
Cost to small refiner buyers	$5,884,765
Small refiner potential buyers	50
Cost to small refiner potential buyers	$25,002,938
Small refiner sellers	26
Value to small refiner sellers	$24,328,990
Large refiners (over 175,000 capacity)	
Total large refiners listed	21
Large refiner buyers	13
Cost to large refiner buyers	$71,262,052
Large refiner sellers	8
Value to large refiner sellers	$40,349,187

Source: U.S. Congress, Senate, Committee on Interior and Insular Affairs, *Hearings on Small Refiners Exemption Act of 1975, S. 861*, 94th Congress, 1st session (1976).

bias to refiners. For 1975, the total subsidy was $251.7 million. In the first six months alone of 1976, this subsidy increased to $211.0 million. As can be seen, this value increases as the size of the refinery decreases.[37]

---

[37] It should be noted that the method of calculating the small refiner bias in fact permits OPEC to calculate the per barrel value of the bias, and thus the amount of cost increase the large refiners pass on to consumers. The value of an entitlement is equal to the difference between the average price of imported crude and the average price of lower tier domestic crude, minus the import fee of twenty-one cents. If OPEC raises its price, the value of an entitlement would increase and, with it, the value of the small refiner bias.

Table 9

VALUE OF SMALL REFINER BIAS TO REFINERS OF VARIOUS SIZES, NOVEMBER 1974–JUNE 1976

($ per barrel)

Month	Refinery Runs (mbd)					
	10	20	50	70	100	150
1974						
November	$ .62	$ .37	$.09	$.07	$.04	$.01
December	.62	.37	.09	.07	.04	.01
1975						
January	.74	.45	.19	.12	.08	.02
February	.84	.50	.21	.14	.08	.02
March	.90	.54	.23	.15	.09	.02
April	.90	.54	.23	.15	.09	.02
May	.91	.55	.23	.15	.10	.02
June	.97	.58	.25	.16	.10	.02
July	1.01	.60	.25	.17	.10	.02
August	1.03	.62	.26	.17	.10	.02
September	1.03	.62	.26	.17	.10	.02
October	1.07	.64	.27	.18	.11	.02
November	1.11	.66	.28	.18	.11	.02
December	1.06	.63	.27	.18	.11	.02
1976						
January	1.00	.60	.25	.17	.10	.02
February	1.81	1.06	.33	.20	.10	.02
March	1.81	1.07	.33	.20	.10	.02
April	1.80	1.06	.33	.20	.10	.02
May	1.80	1.06	.33	.20	.10	.02
June	1.81	1.07	.33	.20	.10	.02

Note: This table ignores special provisions for small refiner entitlement buyers.
Sources: Calculated from 10 C.F.R., Section 211.67(e); and FEA, "Value of Entitlement," *Monthly Energy Review*, September 1976.

Table 11 shows how many months it would take for the value of the entitlements bias to pay for the cost of a new refinery.[38] If the regulatory program were seen as likely to last longer than the next two

---

[38] These figures assume no change in the OPEC price. If the world price for crude rises faster than domestic crude oil prices, the number of months required would fall, because the value of an entitlement, and the bias, would rise.

Table 10

VALUE OF SMALL REFINER BIAS: ENTITLEMENTS, 1975–1976

Month	Number of Entitlements Issued Under Bias	Price of Entitlement	Value of Transfer
1975			
January	3,184,284.98	$6.00	19,105,709.88
February	2,851,873.48	6.75	19,250,145.99
March	2,815,797.62	7.31	20,583,480.60
April	2,708,135.40	7.29	19,742,307.07
May	2,684,602.60	7.39	19,839,213.21
June	2,502,712.00	7.82	19,571,207.84
July	3,053,186.08	8.13	24,822,402.83
August	2,942,345.62	8.31	24,450,892.10
September	2,759,840.55	8.31	22,934,274.97
October	3,145,103.38	8.62	27,110,791.14
November	3,076,268.80	8.94	27,501,843.07
December	3,098,658.00	8.55	26,493,525.90
1976			
January	3,275,125.50	8.09	26,495,765.30
February	3,109,750.25	7.85	24,411,539.46
March	3,368,777.84	7.89	26,579,657.16
April	5,582,095.75	7.85	43,819,451.64
May	5,770,391.42	7.82	45,124,460.90
June	5,640,142.11	7.91	44,613,524.09

Sources: FEA, Entitlement Program Data Base Summary (internal document); and FEA, *Monthly Energy Review*, September 1976.

and a half years, some potential refiners might consider it worth their while to build small refineries based on the returns expected from this bias alone.

The product mix produced from new capacity built in response to regulatory incentives would not be the same as that most desired by society; the result could be shortages during the period of regulations and unnecessarily higher prices after regulations end. Refineries, when built, are designed to yield a particular mix of products. Though this mix can be altered, drastic changes in it or the addition of new products to the total output often require adding new equipment. The cost passthrough regulations induce substitution of other inputs for capital

Table 11

REQUIRED NUMBER OF MONTHS FOR THE ENTITLEMENTS RECEIVED DUE TO SMALL REFINERS BIAS TO PAY FOR A NEW REFINERY OF DIFFERENT CAPACITY

(value of entitlement = $7.91)

Size of Refinery (mbd)	Cost per Barrel of Daily Capacity				
	$500	$600	$700	$800	$900
0–10	9.2	11.1	12.9	14.7	16.6
20	15.6	18.7	21.8	24.9	28.0
30	20.1	24.1	28.2	32.2	36.2
40	32.4	38.9	45.4	51.9	58.3
50	50.7	60.8	70.9	81.1	91.2

Source: Calculated from Table 9.

wherever possible. New equipment, however, is what is needed to give most refineries additional capability to vary their output mix.

The incentive to build small refineries instead of large ones worsens the problem of disincentives to add capital. The small refiner biases of the regulations encourage excessive use of small refineries and make the industry less able to alter the product mix desired at the lowest possible cost achievable with economies of scale in refinery equipment.

Table 11 shows that the lower the costs per barrel of daily capacity, the lower is the time required for the entitlements bias to repay the cost of building the refinery, if the small refiner can keep the entire subsidy—as he can with product decontrol. Costs per barrel of daily capacity for any given refinery size are minimized when a refinery is constructed with the capability for only very simple processes. This relationship is shown in Figure 1, in which complexity is a measure of the kinds of processes installed.[39] The regulations bias the choice of new entrants into refining not only towards small refineries but also towards the construction of simple refineries that would be designed to turn out almost exclusively residual oil and distillates, rather than the various grades of gasoline, petrochemical feedstocks, and the like.

---

[39] For a detailed explanation of the complexity factors of various processes, see the series by W. L. Nelson in the *Oil and Gas Journal*, the issues of September 13, 20, and 27, 1976.

Figure 1

APPROXIMATE COST OF PROCESS UNITS AT U.S. GULF COAST REFINERIES

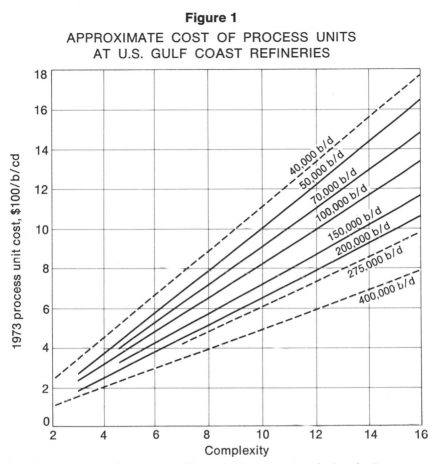

Note: Dashed lines indicate pattern if complexity concept worked perfectly.
Source: *Oil and Gas Journal*, September 20, 1976, pp. 202-203.

A tendency towards the construction of smaller, less complex refineries and of less total capacity can be seen in the statistics of the last few years. Table 12 shows the changes in total U.S. refinery capacity from January 1 to January 1 for the years 1973–1976, in terms of both crude oil refining capacity and vacuum distillation capacity. Vacuum distillation is an early step in producing most refined petroleum products. If vacuum distillation capacity is the last process used on crude oil, refiners will produce a higher proportion of residual fuel oil and of distillates. An increase in the proportion of gasoline output requires other kinds of processing. As can be seen from the figures, the annual rate of growth of vacuum distillation capacity increased between 1973 and 1976; but the rate of growth of

Table 12
ANNUAL CHANGES IN REFINING CAPACITY, 1973–1976

	Percent Change		
	1973–1974	1974–1975	1975–1976
Crude refining capacity	6.3	3.9	1.4
Vacuum distillation	2.9	3.7	3.2

Source: Calculated from data given in the Annual Refinery Issues of *Oil and Gas Journal*, 1973-1976.

total refining capacity slowed down. The jump in crude oil prices and the worldwide recession, both of which lowered demand for refined products and resulted in widespread idle capacity, explain the slowdown in the growth of total refining capacity. The regulations of this period created incentives which explain, in part, the difference in the trends revealed in Table 12.

First, regulations froze crude oil prices at levels lower than world oil prices in 1973. This gave domestic refiners with access to price-controlled oil an advantage over both those refining imported oil and those importing refined products. FEA established the entitlements program on crude oil at the end of 1974, giving all domestic refiners an advantage over all importers of refined products. The major imported refined product was residual fuel oil, particularly to the East Coast. Before the establishment of the price control program, imports of residual fuel oil cost less than domestically refined supplies because of both lower crude oil costs abroad and the high cost of transporting domestic supplies overland from refineries to the market, as residual fuel oil is very viscous and, hence, is difficult to ship through pipelines. Shipping it by tanker is the least cost method of transportation. Imports were, therefore, encouraged until the inception of controls. The controls themselves thus encouraged domestic production of residual fuel oil.

In addition, the original price control regulations gave refiners more latitude in pricing residual fuel oil than other products, so that its price rose to the world level. This altered the relative prices of the various refined products, and it also contributed to the increasing attractiveness to domestic refiners of producing residual fuel oil.

The statistics shown in Table 12 are consistent with a change in refinery capacity designed to alter the mix of those products in such a way as to increase the relative output of residual fuel oil. Such a

Table 13

CONSUMPTION, PRODUCTION, AND IMPORTS OF REFINED PRODUCTS, 1973–1976
(thousands of barrels per day)

	Domestic Demand	Production	Imports
1973			
Total refined products	17,308	N.A.	3,012
Residual fuel oil	2,822	971	1,853
Distillate fuel oil	3,092	2,820	392
Jet fuel	1,059	859	212
Motor gasoline	6,674	6,527	132
1974			
Total refined products	16,653	N.A.	2,635
Residual fuel oll	2,639	1,070	1,587
Distillate fuel oil	2,948	2,668	289
Jet fuel	993	836	163
Motor gasoline	6,537	6,358	204
1975			
Total refined products	16,291	N.A.	1,888
Residual fuel oil	2,433	1,235	1,194
Distillate fuel oil	2,849	2,653	153
Jet fuel	1,001	871	133
Motor gasoline	6,674	6,518	184
January–June 1976			
Total refined products	16,982	N.A.	1,747
Residual fuel oil	2,487	1,300	1,150
Distillate fuel oil	3,130	2,755	117
Jet fuel	983	919	93
Motor gasoline	6,836	6,690	104

Note: 1973-1975 data from Bureau of Mines; January-June 1976 data from American Petroleum Institute.

Source: FEA, *Monthly Energy Review*, August 1976, pp. 8-16.

shift is also reflected in the statistics on actual production and consumption of refined products, as shown in Table 13.

Over time, the construction of less complex, small refineries will do more than simply increase the relative production of those products that require less processing. If the trend towards less complex and small refineries continues, future refinery stock will be less able to respond to changes in the product mix demanded without incurring

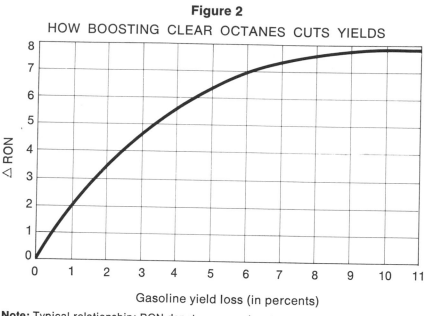

Figure 2

HOW BOOSTING CLEAR OCTANES CUTS YIELDS

Gasoline yield loss (in percents)

Note: Typical relationship; RON denotes research octane number.
Source: *Oil and Gas Journal*, July 26, 1976, p. 72.

costs that are higher than would otherwise have been necessary. Because flexibility will diminish, price changes in response to a given change in demand will be greater than under less rigid conditions.

Changes in conditions and technology are often responsible for changes in the desired output mix from refineries. For example, the conversion of the airline fleet from propeller-driven craft to jets has increased the demand for the various types of jet fuels relative to the demand for aviation gasoline. Similarly, the desire of Congress both to clean up air quality and to improve the gasoline mileage of automobiles has led to the use of the catalytic converter which requires the use of unleaded gasoline instead of leaded gasoline. The desire of the Environmental Protection Agency (EPA) to reduce the lead content of gasoline going into all cars adds to the nature of this underlying shift in demand.

The problems the original EPA lead phasedown regulations have posed for the refining industry serve as an illustration of the way FEA regulations will make response to such demand changes more difficult. Lead compounds are additives that boost the octane rating of gasoline. If lead is not used, either more of the original barrel of crude oil or more processing is needed to make the same quantity of

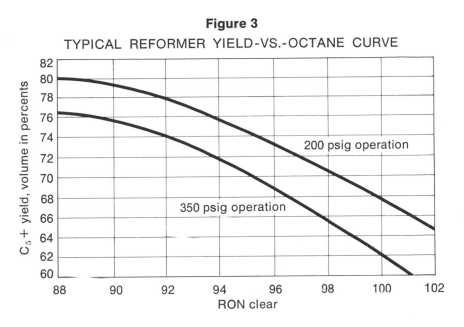

Figure 3

TYPICAL REFORMER YIELD-VS.-OCTANE CURVE

200 psig operation

350 psig operation

Note: For 130-310°F. Arabian feed. RON denotes research octane number.
Source: *Oil and Gas Journal*, July 26, 1976, p. 72.

gasoline of the octane rating that the use of lead would allow.[40] Crude oil used in making gasoline is not available for use in making other refined petroleum products. Further processing to produce clear octanes is a substitute for lead in boosting the octane level of gasoline, and the increased use of clear octanes in gasoline reduces the total amount of gasoline available from a barrel of crude oil. Figure 2 shows this relationship for the refining industry as a whole. The problem is aggravated if the equipment installed in a refinery is of an older type which operates at a high pressure, as shown in Figure 3. Thirty-two percent of existing reforming capacity is of this older type.[41]

Refiners have responded to the increased demand for unleaded gasoline occasioned by the use of the catalytic converter by increasing the lead content of regular gasoline in order to free more clear octanes for the production of unleaded gasoline. Overall, however, the average

---

[40] Refiners are now beginning to substitute manganese for lead, but this is not a fully developed process and also would be likely to require some investment in new equipment if total output of refined products is to remain unchanged. Thus, the possibility of complying with the original EPA lead regulations by the substitution of manganese will not be discussed here.

[41] Leo R. Aalund, "Lead Stepdown Threatens '77 U.S. Gasoline Crunch," *Oil and Gas Journal*, July 26, 1976, p. 73.

lead content of the total gasoline pool has declined since 1970, when it averaged 2.4 grams per gallon, to an estimated 1.9 grams per gallon in July 1976.[42] The original phasedown regulations would have required refiners to reduce this average to 1.4 grams per gallon by October 1, 1976. One refiner estimated that this would reduce the ability of the nation's refiners to produce gasoline by 800,000 barrels per day, or approximately 11 percent of June 1976 production.[43] These numbers are for the refining industry as a whole. Small refiners, with less complex refineries, would be hit harder than large ones. According to a survey of refiners with less than 50,000 barrels-per-day capacity conducted by the American Petroleum Refiners Association, phasedown to an average lead content of 1.5 grams per gallon (0.1 grams per gallon more than the original EPA proposal) would cut gasoline production by those responding by an average of 77 percent.[44]

The new EPA lead phasedown rules require no change in average lead content until January 1, 1978, when the lead content must average no more than 0.8 grams per gallon. By October 1, 1979, the average lead content must be reduced still further to 0.5 grams per gallon.[45] Moreover, small refiners may be subject to different treatment, a possibility still under study by EPA. One reason why EPA gave more time to reduce lead content was that the agency estimated gasoline shortages during the summer of 1977 of up to 500,000 barrels per day and up to 700,000 during the peak driving season of the summer of 1978.[46]

FEA regulations will work to increase the likelihood and the magnitude of shortages due to shifts in demand like that caused by EPA lead phasedown regulations. This occurs because the FEA regulations discourage overall capital investment in refining, and they bias the investment that does occur towards the construction of refineries that either are too small to produce the demanded products at lowest cost or are designed to produce less complex refined products. The result will be shortages during the period of FEA regulation and higher prices after FEA regulations expire.

Entitlements on crude oil alone will not serve as an incentive to expand capacity. Entitlements on crude oil alone (with the exception of the partial entitlements for residual fuel oil) provide a large barrier

---

[42] Ibid., pp. 72-73.

[43] Ibid., p. 73; June 1976 production figure from FEA, *Monthly Energy Review*, August 1976, p. 10.

[44] *Oil and Gas Journal*, July 26, 1976, p. 74.

[45] *Oil and Gas Journal*, September 27, 1976, p. 25.

[46] Ibid.

against the importation of refined products. In 1975, the difference in crude oil costs between U.S. and foreign refiners was approximately $3.50 per barrel.[47] Such a large cost advantage might be thought to provide an incentive to expand domestic refining capacity. Entitlements will end, however, with the end of crude oil price controls in 1979. Thus, they serve only as a temporary barrier to foreign imports. Because refineries have a lifetime of twenty years or more, protection that would serve as an incentive to build new capacity would have to last roughly as long as the life of the refinery.[48] Thus, the entitlements program provides a strong incentive only to expand utilization of existing capacity, as will be discussed in more detail below.

The Barrier to Foreign Competition Provided by the Entitlements Program Will Ultimately Raise Costs and Consumer Prices of Domestically Refined Products. The following two points illustrate how the present FEA regulations cause higher costs in the production of refined products.

Because of the crude entitlements program, domestic refineries are run at higher utilization levels than foreign refineries, and the domestic utilization levels will reach levels where out-of-pocket costs for producing additional units of output rise sharply, raising consumer prices. The entitlements program, because it applies mainly to crude oil imports, provides an incentive to substitute domestic refining of imported crude for product imports.[49] Domestically refined products are sold for prices that reflect the cost of the last barrel of oil used. Given entitlements, the cost of the last barrel of crude oil to the refiner is roughly the average cost of all oil used in the country, rather than the full import price. Given workable competition in the refining sector, the price charged for refined products reflects this average crude oil cost, while imports of refined products reflect the world crude oil cost, which is presently higher by about three dollars per barrel. Because domestic refiners enjoy this crude oil cost advantage, product imports can compete with domestic production only if imports enjoy equivalent production-cost or transportation advantages. Since foreign products generally do not have such large cost advantages,

---

[47] FEA, *Monthly Energy Review*, July 1976, p. 64.

[48] This is the kind of protection that can be provided through tariffs and fees.

[49] This incentive is recognized in FEA, *Preliminary Findings and Views Concerning the Exemption of Motor Gasoline from the Mandatory Allocation and Price Regulations*, November 16, 1976, p. 29 (hereinafter, *Gasoline Decontrol Report*). There are, however, some partial entitlements for imports of residual fuel oil.

despite their currently depressed prices, domestic refiners can enter a market, undersell foreign products, and still earn profits.[50]

Continued substitution will push the level of domestic refinery utilization into the range in which the out-of-pocket cost of an additional unit of output increases. At any given time, increasing the quantity of output produced means raising the level of utilization of existing refineries, as new refineries take time to build. For utilization rates up to about 82–85 percent of capacity, the additional cost imposed by increasing output by one unit remains constant. That is, changing output by one barrel per day at 70 percent of capacity involves the same out-of-pocket costs as does increasing output by one barrel per day when the refinery is operating at 60 percent of capacity. When the level of utilization reaches about 82–85 percent, however, the additional costs imposed by increasing output by a barrel per day begin to rise and, at some point, they rise quite substantially.[51]

As noted above, rising utilization rates, normally a signal for capacity expansion, are not likely to be met by adequate capacity expansion under the present regulations. Instead, as demand increases, U.S. refineries will operate at levels where out-of-pocket costs for an additional unit of output are greater than at present. As the out-of-pocket costs rise, so will prices to the consumer.[52] There is nothing inherently inefficient with refineries operating in a range where incremental costs are rising. In fact, in the long run, it is generally necessary to operate in such a range in order to earn a reasonable return on

---

[50] FEA data show that in 1976, while demand for gasoline increased, imports of gasoline decreased. At the same time there has been a substantial increase in imports of crude oil. FEA states that the reduction in imports "occurred during a period of record demand indicating the ability of domestic refinery production to meet that demand." (*Gasoline Decontrol Report*, p. 25.) As more fully developed below, the task force does not agree with this conclusion. Rather, the decline in imports during record demand appears to indicate the ability of the present regulatory structure to protect the domestic refining industry from foreign competition.

[51] For the detailed measures of the cost functions, see J. M. Griffin, "The Process Analysis Alternative to Statistical Cost Functions: An Application to Petroleum Refining," *American Economic Review*, vol. 62 (March 1972), pp. 46–56; and J. M. Griffin, *Economic Capacity in the Joint Product Case: The United States Petroleum Refining Industry* (Lexington, Massachusetts: D. C. Heath & Co., 1971).

[52] The market sets the price at the level that just covers the out-of-pocket costs of producing the last unit of output demanded. If a refiner finds that the out-of-pocket cost of an additional unit of output is less than the market price, then he will increase his profits by producing that additional unit. On the other hand, if the out-of-pocket cost is greater than the market price, the refiner would lose money by producing that additional unit. The profit-maximizing level of output is where the out-of-pocket cost of an additional unit of output just equals, but does not exceed, the price obtainable.

Table 14

U.S. REFINERY UTILIZATION, 1975–1976

(refinery runs as a percent of rated capacity)

Month	Monthly Average 1975	Monthly Average 1976
January	86.2	85.9
February	83.8	88.7
March	82.2	87.9
April	82.2	87.0
May	82.5	86.5
June	84.5	93.2
July	88.9	91.9
August	89.6	91.1
September	89.2	90.2
October	85.9	86.7
November	87.1	90.3
December	88.8	—

Source: American Petroleum Institute, Weekly Oil Statistics, published in *The Wall Street Journal*, Commodities Section, usually on Thursday of each week.

invested capital. Consumers lose, however, when domestic refineries operate in a range of increasing incremental costs, while foreign refineries, because of excess capacity, operate at lower utilization with lower incremental cost. In this situation, the U.S. economy would gain by shifting some purchases from high-incremental-cost domestic refineries to low-incremental-cost foreign refineries.

U.S. capacity utilization is near the point where out-of-pocket costs rise for additional output. For the week ending November 19, 1976, refineries operated at 90.4 percent of capacity and for the week ending November 26, 1976, utilization was 93.3 percent. During the first ten months of 1976, utilization averaged 88.8 percent of capacity; in the comparable period in 1975, utilization averaged 85.5 percent of capacity. The figures for 1976 are shown in Table 14. Of particular note are the figures for June through September 1976 when utilization exceeded 90 percent of capacity. Although 1976 was a period of record demand, it was also a period in which imports of gasoline, at least, declined. The high utilization levels in this period, along with high refiner price/cost margins, provide further evidence of the way in which the present regulatory structure insulates domestic refiners from foreign competition.[53]

A sharp contrast to these U.S. utilization figures is seen in the utilization rates of European and Caribbean refineries shown in Table 15. In 1975, while U.S. refineries were operating near 86 percent of capacity, European and Caribbean refineries operated at or near 65 percent of capacity.[54] The wide divergence between U.S. and foreign utilization rates during the recent recession suggests that the price effect described above may have occurred. If the regulatory program placed product imports on a more equal footing with crude oil imports, the gap between domestic and foreign refinery utilization would narrow. The U.S. economy would benefit by purchasing slightly more product from underutilized foreign refineries, while at the same time lowering the incremental cost, and price, of domestically refined products.

The economic gain from reallocating some use of petroleum products to imports is not limited just to the present period of depressed demand. The barrier to imports provided by the current entitlements program will allow domestic refiners to meet increases in demand, projected by FEA to grow by 3 to 5 percent annually between 1976 and 1978, by increasing domestic utilization instead of allowing imports to meet some of the increased demand.[55] This increased utilization without capacity expansion, which is not encouraged by the present regulatory system, will eventually mean higher prices for petroleum products in the United States.

---

[53] FEA data show that the refiner/distributor margin (cents per gallon) for motor gasoline varied as follows during the years preceding 1976: 9.8 (1970), 9.5 (1971), 9.2 (1972), 9.7 (1973), 9.3 (1974), and 10.9 (1975). During 1976 the margin increased as follows: 11.5 (January), 11.7 (February), 11.7 (March), 11.2 (April), 12.5 (May), 14.1 (June), 14.6 (July), and 14.5 (August). (*Gasoline Decontrol Report*, p. 40.) Note the correlation between the high margins in June through August with the high utilization levels for these months shown in Table 4. FEA data also show that "recent overall industry rates of return are higher than the historical rates for the petroleum industry." *Gasoline Decontrol Report*, pp. 73 and 127.

[54] A similar contrast is noted by FEA in *Gasoline Decontrol Report*, p. 97.

[55] FEA, *National Petroleum Supply and Demand, 1976-1978*, May 1976, p. 3. FEA recognizes that if demand grows near the rate of 5 percent through 1978, projected utilization rates near 94 percent of capacity will be required if there is no increase in imports. FEA also recognizes that such high rates of utilization can be maintained for no more than a few months. (*Gasoline Decontrol Report*, pp. 78 and 85.) Thus, FEA's upper demand projections are realized, imports will occur anyway but at much higher prices than if product entitlements were granted. In fact, FEA appears to view excess foreign refinery capacity and the possibility of imports as a safety valve available in the event its highest demand projections materialize. (*Gasoline Decontrol Report*, pp. 98-99.)

Table 15

FREE WORLD REFINERY CAPACITY, 1973–1976

	1973		1974		1975		1976
	Refinery capacity (thousand b/d)	Refinery utilization (percent)	Refinery capacity (thousand b/d)	Refinery utilization (percent)	Refinery capacity (thousand b/d)	Refinery utilization (percent)	Refinery capacity (thousand b/d)
United States	13,454	92	14,220	89	14,961	86	15,237
Japan	4,526	98	5,152	83	5,346	76	5,860
Canada [a]	1,725	98	1,788	99	1,878	91	2,024
Caribbean	3,779	97	4,336	71	4,267	65	4,273
Middle East	2,758	91	2,882	86	3,281	83	3,285
Other	6,519	91	6,966	83	7,781	73	8,747
Total Western Europe Includes:[b]	16,827	88	18,110	76	18,718	65	19,972
West Germany	2,698	86	2,826	76	2,987	62	3,103
France	2,949	92	3,140	82	3,342	65	3,312
United Kingdom	2,465	93	2,762	81	2,783	67	2,889
Italy	3,593	73	3,882	62	3,953	51	4,082
Netherlands	1,826	79	1,826	71	1,841	62	1,985
Total	49,588	92	53,454	82	56,232	75	59,398

[a] Canada also subsidizes crude oil imports.

[b] Only selected countries of Western Europe are included here. The total for these countries does not equal the total for all of Western Europe.

Note: The figures are as of the beginning of the year.

Source: Central Intelligence Agency International, *Free World Refining Industry Plagued by Surplus Capacity* (unclassified), July 29, 1976.

The small refiner biases in FEA programs encourage the use of smaller, more costly refineries at the expense of larger, less costly refineries, raising consumer prices. The workings of several FEA programs encourage the use of small refineries at the expense of larger ones. The most important biases in favor of small refiners are found in the entitlements program; others exist in the price control regulations and the crude oil buy/sell program.

The small refiner biases have been effective in changing the proportion of the market served by small refiners. In 1972, refiners with less than 175,000 barrels-per-day capacity accounted for 14.4 percent of total refined product sales. In 1975, they accounted for 17.8 percent.[56] Moreover, examination of the types of equipment installed in refineries of various sizes shows that small refiners are not overwhelmingly producers of specialty products or even of just residual and distillate fuel oils.[57]

The announced purpose of the small refiner biases is to maintain the competitive viability of small refiners by continuing a subsidy which began under the Mandatory Oil Imports Program (MOIP), but which would otherwise have ended with that program. Presumably, the decision to retain a bias in favor of small refiners was made because they face higher operating costs at any given level of utilization than larger refineries do. It is interesting to note, however, that the value of the bias under the MOIP amounted to approximately $0.15 per barrel at its height for refiners with throughput of 10,000 barrels per day in contrast to up to $1.80 per barrel, as shown in Table 9.[58]

The small refiner biases built into the entitlements program amount to a subsidy of small refiners by the larger ones. Larger refiners pay this subsidy in the form of higher crude oil costs.[59] These higher crude oil costs are, in turn, passed on to the consumer.

---

[56] Calvin T. Roush, Jr., "Crude Oil Allocation and Refiner Market Shares," paper presented at the American Economics Association Meeting, September 18, 1976.

[57] Bureau of Mines, "Petroleum Refineries in the United States and Puerto Rico," *Petroleum Refineries, Annual,* January 1976.

[58] Roush, "Crude Oil Allocation," pp. 24-25. The $0.15 value is for 1969 when Roush states that quota tickets were at or near the highest value ever attained. See also Kenneth W. Dam, "Implementation of Import Quotas: The Case of Oil," *Journal of Law and Economics,* vol. 14 (1971), p. 23.

[59] This argument applies most strongly when the output mix from large and small refineries is roughly the same. Many small refineries in the United States do produce a wide range of outputs. One test of the complexity of the refinery, and therefore its ability to produce a wide array of products, is whether it produces gasoline. A survey of forty-three refineries of less than 50,000 barrels-per-day capacity showed that twenty-five produced gasoline along with other products, eighteen did not. (*Oil and Gas Journal,* July 26, 1976, p. 74.) A similar conclusion is shown by the data in Bureau of Mines, "Petroleum Refineries in

Refiners with less than 30,000 barrels-per-day capacity accounted for only 5.9 percent of total domestic refinery capacity in January 1975.[60] They are thus not likely, in the short run, to be the sources of additional supplies; the large refiners fill this role. The market price in turn reflects the out-of-pocket costs of the last barrel of output produced. The subsidy to small refiners is paid by the large refiners which produce that last barrel of output. The subsidy increases the out-of-pocket cost of an additional barrel of oil for large refiners because it reduces the fraction of an entitlement which a large refiner receives when he processes an additional barrel of imported crude oil. The more entitlements given to small refiners, the smaller is this fraction (the NDCOSR), and thus the greater is the effective cost of an additional barrel of crude oil. Moreover, the amount of the subsidy that any given large refiner pays varies slightly with the amount of crude oil that that large refiner processes. The variation occurs because any increase in the amount of imported crude oil processed decreases the NDCOSR and increases slightly the proportion of entitlements that the refiner is allocated. Again, because the large refiners are the providers of expanded levels of output, an increase in their out-of-pocket costs means a rise in the market price. The market price level is thus likely to be higher than it would be in the absence of the subsidy.

Under the price control rules for products that still are controlled, recipients of the subsidy must reflect any such crude oil cost savings in prices. To the extent that the savings push small refiners' prices below market levels, they may be able to use accumulated banks to justify keeping prices at market levels despite the more favorable crude oil costs. However, for products that have been decontrolled, small refiners need no banks to keep their prices at the market levels. Thus, the proportion of the subsidy that is allocated to middle distillates, residual fuel oil, and naphtha jet fuel can be kept by small refiners.[61] In any event, it is the belief of Gorman Smith, assistant

---

the United States and Puerto Rico." Some small refineries, however, specialize in certain products for which demand is low or for which the economies of scale in production are less. Moreover, the minimum optimal size for a refinery engaging in no further downstream processing than vacuum distillation (and thus producing mainly residual and distillate fuel oils) is less than for a refinery equipped to produce a wider range of products.

[60] Refiners with less than 10,000 barrels-per-day capacity accounted for 1.4 percent of total capacity. S. Rep. No. 94-1005, 94th Congress, 2d session (1976), p. 49.

[61] The result of this may be to induce small refiners to increase the proportions of distillates and residual fuel oil that they refine in order to retain more subsidy benefits for themselves.

administrator for regulatory programs, that small refiners are charging the market price for their products.[62]

The distortion created by the small refiner bias may tend to worsen in the long run. Entitlements are granted on the basis of a company's total refinery throughput, regardless of whether it operates one or several refineries. Because the small refiner bias is greater at low throughputs, there is a strong incentive for multirefinery companies to sell their smallest refineries to companies that own no other refineries. This would increase the number of entitlements that go to small refiners and thus raise the effective cost of crude oil to the larger companies, since the number of entitlements is fixed at any point in time. Again, as the market price is determined by the larger refiners' costs, because they are the source for additional output, this incentive to spin off small refineries can raise product prices.

Other provisions of the FEA regulations reinforce the incentive, inherent in the small refiner biases of the entitlements program, to use small refineries before larger ones. The class-of-purchaser rules that limit the variation of prices within regions favor small refiners who are likely to serve only one market. Refiners may pass on increases in the cost of crude and selected nonproduct inputs and, if market conditions do not permit the passthrough, refiners may bank the cost increases for later passthrough. However, a refiner must charge the same amount to all members of a class of purchaser, nationally for all but gasoline, and within a PAD district for gasoline.[63] The prices charged for gasoline to members of the same class of purchaser in different PAD districts can vary by no more than three cents. Moreover, a refiner can only bank the increase in costs not passed through as if all classes of purchasers received the same cost increase as the greatest amount actually passed through. These two rules put refiners serving more than one market within a PAD, or more than one class of purchasers, at a distinct disadvantage in comparison with a small refiner serving only one market and one class. Because the ability of different markets to absorb price increases differs, larger refiners will have to absorb some cost increases in markets where demand is weak, while not being able to take full advantage of strong demand in other markets. Refiners with more limited markets, on the other hand, can either take full advantage of the strong market conditions or bank the full increase in costs.

The crude buy/sell program also increases the subsidy paid to small refiners by the major refiners. Only the fifteen largest refiners

---

[62] Interview with Gorman Smith, October 13, 1976.
[63] The United States is divided into five PAD districts.

must sell crude under this program; independent and small refiners are the major recipients of crude oil. The program requires sellers to transfer a quantity of crude oil to eligible buyers. The buyers pay the seller's average price for imports for the oil so transferred. Any difference in the replacement cost and the cost charged to the buyers is added to the product costs of the sellers for the period in which the sales took place. Thus, the cost of buying crude oil for independent and small refinery operations shifts to the large integrated refiners, thereby increasing the crude oil costs paid by them.

FEA Regulations Also Cause Higher Costs in Marketing, Causing Consumer Prices in Turn To Be Higher. The regulatory program of FEA covers not only the production but also the distribution of refined products. Though each refined product is marketed in a different fashion, the marketing of gasoline can serve as an example of the general nature of the effects of the regulatory program.

We recognize that our focus on gasoline creates a very imperfect analogy to some products, especially natural gas liquids. Gasoline is, however, the major product produced, and the study of gasoline marketing does indicate what kinds of problems are likely to arise in the marketing of most refined products.

Consumers are denied the full benefits of economies of scale and specialization in gasoline marketing. For several decades before the Arab oil embargo, the marketing of gasoline was undergoing basic changes. The role of the corner gasoline station that sold branded gasoline and performed a variety of repair services, the so-called full-service station, was losing ground to nonbranded stations that sold only gasoline, and sold it for less than the branded dealers were charging. These nonbranded dealers made their profits by selling higher volumes of gasoline rather than by combining gasoline sales with repair work. Consumers were attracted to such stations by the lower prices. At the same time, the increasing complexity of cars along with the decreasing need for routine repairs and maintenance reduced the need for the nongasoline services of the corner stations.

In addition to pioneering the development of gas-only, high-volume outlets, the nonbranded independents have introduced other innovations in marketing, such as self-service islands. These have worked to reduce further the costs of marketing gasoline, and to help keep prices to consumers lower than they otherwise would have been.

The combination of the freeze on supplier/purchaser relations and the class-of-purchaser rules have worked to continue these trends as revealed in data on market shares shown in Table 16. The figures

Table 16

MOTOR GASOLINE DISTRIBUTED THROUGH JOBBERS AND
DISTRIBUTED DIRECTLY BY REFINERS, 1972–1976

(in percent)

	Sold through Jobbers		Direct Refiner Distribution				
Year	Branded	Non-branded	Refiner operated	Branded lessee dealers	Branded open dealers	Bulk purchaser consumers	Total
1972	20.4	15.2	7.8	36.6	11.1	8.9	100.0
1973	21.1	13.4	8.7	36.7	11.7	8.4	100.0
1974	21.3	16.0	9.1	33.7	11.6	8.3	100.0
1975	22.9	18.4	10.4	29.8	10.8	7.7	100.0
1976[a]	24.1	19.1	11.5	28.0	10.4	6.9	100.0

[a] Preliminary figures based on sales through June 1976.

Source: FEA Refiner Survey (FEA P-305-S-0 and P-036-M-0), as cited in FEA, *Preliminary Findings and Views Concerning the Exemption of Motor Gasoline from the Mandatory Allocation and Price Regulations*, November 16, 1976, p. 58.

indicate that market shares of both refiner-owned and jobber-supplied stations have increased, while the share of branded independent dealers has decreased. This is the likely result, however, of the combination of the freeze on supplier/purchaser relations and the class-of-purchaser rules rather than a continuation of cost-cutting innovations. The rules governing the freeze on supplier/purchaser relations, by not requiring a purchaser to purchase his allocation from his base period supplier, unofficially permit the purchaser, but not the seller, to terminate a supply relationship. Thus, refiners have to continue to supply all of their base year customers, plus any others that have been assigned to them by FEA, until such time as those purchasers decide they no longer wish to be supplied by that refiner. At the same time, the class-of-purchaser rules, in conjunction with the banking provisions, force refiners to maintain the price differentials that were in effect during the base period. The result is that nonbranded independents and refiner-owned stations receive gasoline at lower wholesale prices than do the various classes of branded independent stations. These provisions force gasoline supplies to the former two types of stations, not because they are necessarily more efficient at selling it, but because they had received favored treatment in the past.

The allocation regulations compound the effects created by the freeze on supplier/purchaser relations. Under the procedures for granting new entrants base period allocations, new stations receive an allocation based on a comparison to the allocation entitlement of a similar station in the market area, whereas old stations have allocations based on their operations during the base year. Because allocations serve as a form of insurance in case of another embargo, the discrepancy in allocations, based not on performance but on year of entry, gives an advantage to new stations. This advantage, in turn, reduces the incentive for a marketer to expand an existing station, because it will be more difficult to increase its allocation than it will be to get an allocation for a new station. Thus, instead of enlarging and consolidating existing stations, marketers will abandon them and start over or just expand the number of stations. Any higher costs created by these regulations are passed on to consumers in the form of higher prices.

Supplies are prevented from moving easily into regions of growing demand. Regulations forbid refiners to charge different prices to the same class of purchaser within a PAD district and permit only a narrow range of price variation between PAD districts. Refiners therefore find it difficult to respond to varying demand conditions around the country, as indicated by the predictions of localized shortages that appear periodically.[64] If there were no class-of-purchaser restrictions, growing demand in one area would induce a price increase that in turn would motivate suppliers to increase the supplies sent to the region in question. The regulations do not allow prices to serve this function, making it difficult for refiners to respond. This difficulty is compounded when increasing the supplies to a particular region increases costs as well. Thus, if the growing markets are in areas with higher than average costs of transporting products, they will be hindered in obtaining products, so that the probability of experiencing spot shortages increases.

As a Result of FEA Regulations, Wholesale Prices Are Higher than They Would Be if All the Savings from Crude Oil Controls Were Being Passed on to Consumers, and This Difference May Even Increase. The entitlements program, by excluding imports of products, more often than not has prevented consumers from benefiting from the lower product prices that would have prevailed if refined products were granted full entitlements.

---

[64] See, for example, the concern over the possibility of spot shortages reported in the *Wall Street Journal*, July 1, 1976, p. 4.

Entitlements are provided to all refiners based solely on domestic crude oil use. Except for some importers of residual fuel oil, importers of refined *products* do not receive entitlements. Imports of refined products also must pay a larger import fee than imports of crude oil. Thus, the regulatory program creates a twofold disadvantage to imports of refined products vis-à-vis domestically refined products. The magnitude of this disadvantage has ranged as high as 5.5 cents per gallon on regular gasoline. Had entitlements been granted to gasoline importers, only the 1.5 cents-per-gallon advantage for domestic products due to the import fee would have remained. More often than not during the period from the establishment of the entitlements program (November 1974) to the present, such a change in the program would have enabled consumers to enjoy lower prices than actually prevailed. Figures 4 to 7 compare the imputed prices if importers of refined products had received entitlements, the actual spot market prices in New York City, and the price in Rotterdam.[65]

The times when the imputed prices of refined products would have been lower had importers of products received entitlements were periods when supplies of products were available abroad at prices which, after the addition of transportation and import fee and subtraction of the value of an entitlement, were below the price of an equivalent domestic cargo in the United States.

Specifically, it appears that:

- The price of regular gasoline would have been reduced throughout most of the period. The amount of reduction would have been from less than one cent to seven cents per gallon (see Figure 4).

- The price of premium gasoline would have been reduced except during the months of April through June 1975 and February through May 1976 (see Figure 5).

- The price of distillate fuel oil would have been reduced during the months of January through May 1975, December 1975, and the first two months of 1976. These are the primary heating months when domestic prices are at their maximum. The magnitude of the reduction would have been two cents to four cents per gallon (see Figure 6).

---

[65] The potential price reduction might be even greater for Caribbean refined products. FEA data indicate that the current landed price of Caribbean gasoline is only slightly above the domestic price. (*Gasoline Decontrol Report*, pp. 128-29.)

- The price of residual fuel oil would have been reduced throughout the period from about one dollar to more than two dollars per barrel (see Figure 7).

These savings could have been, but were not, passed on to consumers if entitlements were extended to importers.

Because of the size of the barrier created by the entitlements program applying basically only to crude oil, the cost effects discussed earlier in this section will continue. The rise both in the number of small refiners and in domestic utilization generally will likely increase the out-of-pocket costs and thus the prices of refined products. This in turn will cause the difference between future imputed prices and future domestic prices to increase as well, at least until foreign refineries achieve equal utilization rates. Even then, however, because of the disincentives to additional refinery construction in the United States, the gap between imputed prices and domestic prices might widen still further.

Extending the entitlements program to cover all product imports in addition to crude imports would involve adjusting the NDCOSR to include product imports, and giving to each product importer the number of entitlements equal to the NDCOSR times the number of barrels imported or, as an alternative, the crude equivalent number of barrels.[66] The effect of the extension would be to lower the effective price of a barrel of imported product by the NDCOSR times the price of an entitlement.[67]

Evidence of the effect on prices of having entitlements for imports of products can be seen in the spot market prices of residual fuel oil after partial entitlements were granted for imports of that product. Figure 8 shows the change.

The low capacity utilization rates abroad reinforce the conclusion that in the short run, at least, prices would be likely to fall if there were product entitlements. While U.S. refineries recently increased their utilization levels, foreign refineries decreased theirs. U.S. refinery throughput increased in 1975 by approximately 65 million barrels; world refinery throughput decreased by roughly 800 million barrels.[68]

---

[66] The denominator of the NDCOSR would become total runs to stills in a month plus the number of barrels of all product imports or, for instance, their crude equivalent. The numerator would remain the same as it is presently.

[67] If an importer buys 100 barrels of product (or crude equivalent), he is given NDCOSR times 100 entitlements. If he sells these, he receives NDCOSR times 100 times entitlement price in dollars. Thus, the imported product effectively costs the importer NDCOSR times entitlement price less per barrel than if he received no entitlements.

[68] L. G. Southhard, *Highlight Report: World Petroleum Supply and Demand*, U.S. Bureau of Mines.

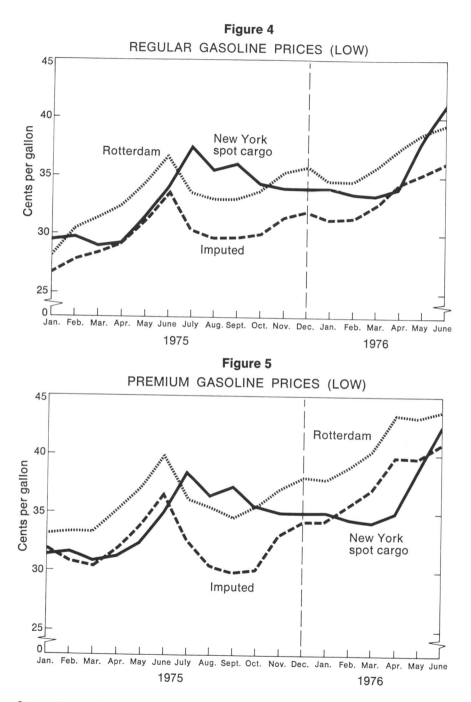

Figure 4

REGULAR GASOLINE PRICES (LOW)

Figure 5

PREMIUM GASOLINE PRICES (LOW)

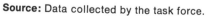

Source: Data collected by the task force.

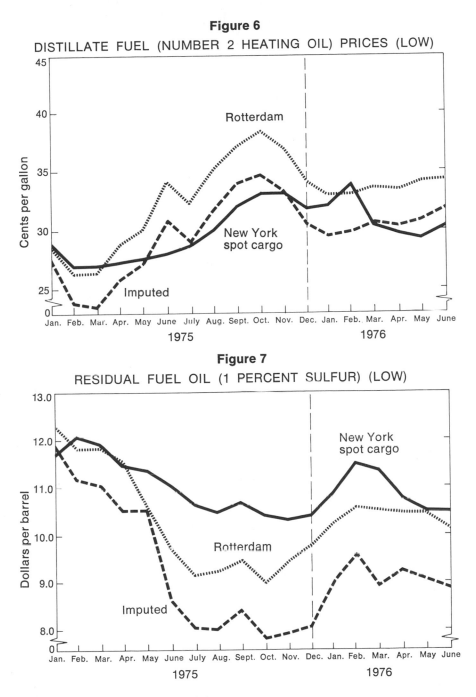

Figure 6

DISTILLATE FUEL (NUMBER 2 HEATING OIL) PRICES (LOW)

Figure 7

RESIDUAL FUEL OIL (1 PERCENT SULFUR) (LOW)

Source: Data collected by the task force.

87

Figure 8

SPOT MARKET QUOTATIONS FOR 1 PERCENT SULFUR NUMBER 6 FUEL OIL

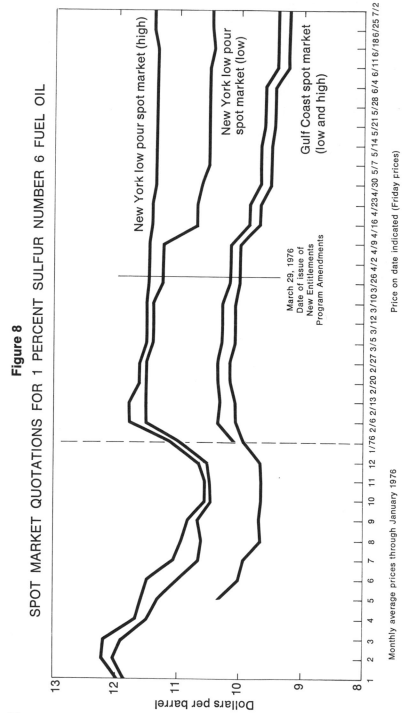

Monthly average prices through January 1976

Price on date indicated (Friday prices)

Source: Calculated from Platt's Oilgram.

The refinery utilization figures in Table 15 show this decrease in throughput.[69]

The low utilization levels shown for most of the rest of the world indicate an ability of the United States to take advantage of bargain prices now, and until such time as world demand pushes foreign utilization levels into the range where out-of-pocket costs of adding more output increase and approach U.S. levels. Because U.S. product imports represent only a very small portion of total world refinery output, adoption of a product entitlements scheme now would not be likely to have an immediate effect on world prices. The most important determinants of world prices will be general economic recovery abroad and the level of OPEC crude oil prices.

---

[69] A similar picture of foreign refinery utilization can be found in the discussions in FEA, *Trends in Refinery Capacity anl Utilization: Petroleum Refineries in the United States; Foreign Refinery Exporting Centers*, June 1976.

3

PROCEDURAL PROBLEMS WITH
THE REGULATORY PROGRAM

The Compliance Program

The purpose of this section of the report is to analyze whether FEA's price and allocation regulations can be effectively enforced in current conditions of adequate supply and in future conditions of severe shortage.

Such an analysis might cover four subject areas: the sufficiency of compliance staff available to FEA; the adequacy of enforcement priorities set by the agency; the enforceability of the regulations according to their relative clarity, simplicity, and conformity with industry record keeping; and the impact of potential sanctions on the industry's will to comply. The task force inquiry focused on the last two issues in reaching the conclusion that the present price and allocation regulations, with the exception of those applying above the refiner level, are inherently too difficult to enforce effectively at an acceptable cost to society. Such difficulty of enforcement is less significant now than may be the case in the event of another embargo: now prices below the refinery gate are constrained by competition and product is in ample supply; but in a future shortage effective enforcement would be necessary to restrain prices and to allocate scarce supplies.

We point out that FEA's position is that it has lacked adequate staff to fulfill its enforcement mission. Although we have no basis upon which to quarrel with this assessment, we do not believe that merely adding personnel will solve the enforcement problems discussed below. Our judgment assumes, moreover, that current and future compliance personnel will be properly utilized.

Historical Overview. The following sections summarize the history of the compliance program during and after the Arab oil embargo.

During the embargo period, FEA's compliance effort (implemented by IRS) was directed toward resolving complaints received from the public regarding the price of petroleum products at the retail level. Upon its establishment in late 1973, the Federal Energy Office (now FEA) and the Internal Revenue Service arranged for IRS to assume responsibility for FEA's enforcement efforts until June 30, 1974.[1] Under the terms of the agreement, IRS immediately detailed 300 auditors and investigators to FEA activities and initiated recruitment and training of 1,000 auditors/investigators for FEA. IRS conducted all field investigative functions in both allocation and pricing.

In the first three months of the agency's existence, FEA (through IRS) received over 128,000 public complaints.[2] Roughly 90 percent of those complaints dealt with *retail* service station "pump prices." IRS enforcement efforts were overwhelmingly related to resolving those complaints and therefore dealt primarily with pricing at the retail levels of the industry. By the end of June 1974, some 16,000 price violations had been uncovered and approximately $30 million in price rollbacks had been achieved.[3]

FEA was also attempting to implement and insure compliance with the new mandatory allocation regulations, a substantial effort in light of the widespread product shortages which had resulted from the embargo and the equally widespread unfamiliarity of the petroleum industry with the terms of the allocation program. Much of the allocation-related enforcement effort was conducted jointly, with operations (FEA) doing essentially "desk audits" of allocation entitlements and compliance (IRS) providing detailed investigative support and remedial functions. Auditor/investigators checked allocation questions as an adjunct of complaint-initiated pricing investigations at the reseller level.

After the embargo, FEA developed a set of enforcement priorities with the refiner audit program being given the highest priority. With the end of the actual embargo in March 1974, the number of public complaints received began to slacken, in part because of increasing awareness within the industry of FEA pricing and allocation require-

---

[1] FEA, *Compliance Manual*, section 1.201 (1975).

[2] U.S. Congress, Senate, *Hearings before the Subcommittee on Administrative Practice and Procedure of the Senate Committee on the Judiciary, FEA Response to Question No. 10,* 94th Congress, 1st session (1975), p. 310.

[3] Memorandum from James R. Newman to John D. Askew, November 17, 1976.

ments, availability of supplies, and public acceptance of higher petroleum prices. After several more months, FEA (which by June had assumed control of its compliance functions from IRS) had to assess its compliance effort and establish its compliance priorities.

The first choice FEA had to make with respect to its compliance effort was between price and allocation. Increased supply levels following the end of the embargo had resolved the major allocation problems. FEA regional and national operations divisions continued to monitor allocation levels and attempted to resolve most of the remaining problems through "jawboning." Thus, the combination of these factors made it logical for FEA compliance to concentrate its efforts in the pricing area, responding to allocation problems only on an "as requested" basis.

After the decision was made to concentrate future compliance efforts on pricing, the next step was to develop price programs and priorities for the various levels of the petroleum industry. The highest priority program was the Refiner Audit and Review Program (RARP). When initiated in January 1974, RARP consisted of identifying the thirty largest refiners and assigning several auditors to each on a continuing basis.[4] Approximately 162 auditors were assigned to refineries as of June 1975. Some 95 auditors were assigned to the thirty major oil companies on a continuing basis, while the rest were in the process of auditing selected small refineries. The first three audit cycles were completed by January 1975. They indicated potential violations totaling $657 million and resulted in $74.8 million in market adjustments in the form of price rollbacks or actual refunds.[5]

Over the next year (1974–1975), several other programs were developed in response to price trends discovered through investigations or in response to congressional concerns.[6] The propane reseller project, the crude oil producer project, and the utility fuels price project, in turn, received high priority attention from FEA compliance.

The propane reseller project grew out of several investigations which disclosed that some propane resellers were conducting "paper transactions" between firms with an apparent goal of increasing prices by "pyramiding" margins on multiple sales of the same product. As

---

[4] FEA, *Compliance Manual*, section 4.200 (1975).

[5] U.S. Congress, House of Representatives, *Hearings before the House Committee on Government Operations*, 94th Congress, 1st session (April 10, 1975), testimony of Gorman Smith; and Senate, *Hearings before the Subcommittee on Administrative Practice and Procedure*, testimony of P. H. Hughes, assistant comptroller general of the United States.

[6] Interview with James Newman, FEA deputy assistant administrator for compliance, September 23, 1976.

of March 24, 1975, there were 108 investigations in progress. Violations were found in 14 cases resulting in refunds of about $4 million and penalties of $22,500. At that time, total violations were eventually expected to be approximately $30 million.[7]

The crude oil producers project was initiated when FEA's monitoring of crude oil production revealed increased volumes of "new" crude in unexpected areas. Investigations were conducted to determine whether producers were incorrectly certifying production for sale at new, that is, uncontrolled, crude prices. There were 125 independent crude oil producers (which reported 25 percent or more of the new crude oil production) selected for investigation to determine how widespread the violations were. By June 1975, 115 investigations were completed with 20 investigations resulting in refunds of $1.1 million and penalties of $59,500.[8]

The utility fuels price project began when utilities reported sharp increases in the cost of residual fuels used at electrical generation plants. In response to concerns expressed by Congress (particularly East Coast delegations), Mr. Zarb ordered a wide-ranging audit of utility prices. While RARP continued to be listed as top priority, personnel were diverted to the utility project for limited periods of time. In some cases, the utility fuels price project was tied to the continuing RARP effort. As of May 16, 1975, there were 336 suppliers of public utilities that had been identified for investigation. Of the 80 cases closed as of that date, 9 suppliers were found in violation, and violations were estimated by the General Accounting Office at $1.7 million, with about $600,000 attributable to violations on sales of fuel oil to other than utilities.[9]

In response to a congressional inquiry, FEA took administrative steps to make RARP a more effective program. In 1975, Senator Edward M. Kennedy (Democrat, Massachusetts) directed an inquiry into FEA's compliance effort.[10] His subcommittee questioned the accuracy of FEA's announced system of priorities and expressed skepticism concerning the effectiveness of several programs, including RARP, on the basis of what it considered insufficient allocation of personnel. Although RARP had been assigned top priority since mid-

---

[7] House of Representatives, *Hearings before the House Committee on Government Operations,* and Senate, *Hearings before the Subcommittee on Administrative Practice and Procedure.*

[8] Ibid.

[9] Ibid.

[10] Senate, *Hearings before the Subcommittee on Administrative Practice and Procedure.*

1974, the audit team sizes at several large refiners were thought to be clearly inadequate.[11] For example, Exxon, which completes some 180,000 transactions per day, was assigned only two or three auditors.[12] Even in instances of teams with adequate staffing, personnel were often diverted to the other high-priority projects such as the utilities project. Thus, it was apparent that the audit force was badly in need of more staff, particularly if the current system of compliance was to be continued for an indefinite period. As a result of the subcommittee's inquiry, FEA instituted a program of quarterly work plans designed to evaluate and revise audit priorities, allocate manpower resources, and monitor progress. Additional auditors were hired and the RARP teams were increased in size.[13] Subsequent to this initial increase, a fiscal year 1976 supplemental budget was prepared, which would bring to 293 the total number of on-site auditors at the major refiners. The actual authorization to hire was not received until August 21, 1976; all 293 were expected to be present by the end of calendar year 1976.[14]

Current Enforcement Problems. For a regulatory program to work, it must be enforceable. The injunctions of the program must be unambiguous and capable of being followed by those covered; violations must be susceptible to timely detection and prompt remedy or sanction; the cost of noncompliance should be greater than the perceived benefits of disregarding the regulations. If regulations are extremely difficult to enforce, assuming reasonable compliance personnel commitments, it probably means the regulations are too complex.[15]

---

[11] Ibid.

[12] James R. Newman, comments at task force staff meeting, August 10, 1976.

[13] FEA, *Progress Report*, presented to Senate Subcommittee on Administrative Practice and Procedure, 1975.

[14] Memorandum from James R. Newman to John D. Askew, November 17, 1976.

[15] The task force notes that a significant portion of FEA's regulatory program, that dealing with the pricing and allocation of crude oil upstream from the refiner level, is enforceable. These producer price regulations are enforceable even though the possibility exists that they may be circumvented by "old/new game playing." (Interview with Gordon Harvey, director of compliance policy and planning, FEA Office of Regulatory Programs.) The probability of detecting violations by producers which falsify records to increase their dollar income from crude production is high. There are 300 first purchasers of crude, 100 of whom purchase 93 percent of domestic crude production. Thus, FEA can effectively monitor the 300 purchasers of crude oil (via a data collection system using P124 Reports) to insure that the composite price (that is, maximum weighted average first-sale price) is correct. For similar reasons, the crude allocation regulations can be effectively enforced.

The complexity and ambiguity of price regulations applying at and below the refiner level make enforcement extremely difficult.

1. Refiner price rules are inherently difficult to enforce: Effective enforcement of the refiner price rules depends upon the ability of the regulated firm to compute and the agency to verify accurately the May 15, 1973, base period price, and then to determine what costs may lawfully be added to that price. Both processes are inherently difficult.

At present, FEA has not completed its verification of the May 15, 1973, base price for major refiners. It was estimated that it would be January 1, 1977, before this task was finished.[16]

The calculations of the May 15, 1973, weighted average selling price (base price), the first component of the maximum lawful selling price, involve three steps. First, all sales on May 15, 1973, must be placed in one of the four product categories. Second, each purchaser of a "covered product" must be categorized and placed into a class of purchaser. Finally, the weighted average selling price for each class of purchaser for all covered products must be computed.

There are two principal reasons why FEA has not yet verified the refiners' May 15, 1973, base price. First, it is necessary to determine the May 15, 1973, price charged for each product for each class of purchaser. This is an enormous task as it can involve auditing the refiners' categories of products and class of purchasers for up to 180,000 daily transactions for a single firm. More importantly, as emphasized by the chronological chart in Table 17, the vagueness of the definition of what constitutes a class of purchaser has resulted in confusion as to what the proper class-of-purchaser determinations are.

Second, in determining the May 15, 1973, weighted average selling price for products, it is necessary to determine when a "transaction" has occurred. FEA has not clarified its definition of transaction. Until transaction is defined, the proper class of purchaser cannot be determined for contract customers, a small but crucial segment of a refiner's customers, which include most utility and some large industrial customers. Some refiners have narrowly interpreted a transaction to occur only on the date when a binding contract is entered into and have excluded, from weighted average selling price computations, sales and deliveries made under contracts existing on May 15, 1973. This interpretation by the refiners has generally resulted in higher May 15, 1973, weighted average prices for some classes of purchasers since, in a rising market, more recent contracts can be expected to

---

[16] Interview with James R. Newman, September 23, 1976.

Table 17

DEFINITION OF "CLASS OF PURCHASER"

Date	FEA Action
January 15, 1974	Initial definition; 10 C.F.R., section 212.31
June 12, 1974	Subsequent clarification; FEA Ruling 1974-17
June 12, 1974	Subsequent clarification; FEA Ruling 1974-18
March 7, 1975	Further clarification; FEA Ruling 1975-2 (takes precedence over FEA Rulings 1974-17 and 1974-18)
April 18, 1975	Instructions to FEA regional staff about what FEA Ruling 1975-2 really meant [a]
October 8, 1976	FEA General Council attempting to define *transaction* [b]

[a] Memorandum from Bob Montgomery and Gorman Smith to FEA regional administrators, April 18, 1975.
[b] Interview with David G. Wilson, principal deputy general counsel, October 8, 1976.

reflect higher prices. Moreover, the fact that FEA has failed to take a firm and clear position regarding the definition of a transaction has created a roadblock to resolving many class-of-purchaser issues.

The agency maintains that it has never had sufficient personnel to audit the myriad transactions which are supposed to reveal the base price. We are uncertain, however, whether reasonable additions to the staff would alleviate the problem. The larger the number of auditors, the greater coverage of firms the agency can achieve. On the other hand, given the changing and ambiguous definitions of transaction and class of purchaser, computation of the base price is not only exceedingly difficult but subject to inherent inaccuracies. The fact that it has required three and one-half years to determine the base price component of the maximum lawful selling price clearly indicates an enforcement problem.

A further element of refiner price regulation compliance is the computation of those costs which may lawfully be added to the base price. The constant adjustments to and interrelationships between refiner cost banks frustrate FEA's compliance effort with respect to the calculation of those increased costs.

Refiners are required to calculate three types of increased costs (crude, purchased product, and nonproduct) for four different time periods for four different categories of products and a separate category for propane. Refiners' normal accounting practices, however, do not keep track of unrecouped costs, even on a product category basis, much less the approximately forty banks required by FEA regulations.[17]

In addition to the complexity of cost-bank accounting and its dissimilarity to general accounting practices, banks must be constantly adjusted to reflect changes in the regulatory program. Refiner cost banks must be adjusted, for example, to reflect the creation of new product categories. More importantly, the banks must be adjusted to reflect the decontrol of products, and FEA has not yet resolved significant problems arising out of product decontrol. As products have been decontrolled, beginning on July 1, 1976, the agency has specifically required the elimination of the banks for that product and has generally prevented the subsequent transfer of any part of that bank to any other product still subject to controls. Prior to decontrol, however, what is known as the "H" factor permitted firms to reallocate or shift increased costs from certain categories of products to motor gasoline for recoupment through gasoline price increases.[18] Thus, some costs in banks allocated to gasoline were originally incurred with respect to other products which have now been decontrolled. FEA has as yet taken no position on whether those costs must be traced down and removed from the motor gasoline banks.

Finally, banks are cumulative and interrelated. An adjustment in one bank for one month, therefore, may affect all the other bank calculations for subsequent months. Because of this interdependency, they will continually be in a state of adjustment, making final enforcement of cost increases extremely difficult.

The complexity of the banking regulations is illustrated by a dispute involving the recoupment of nonproduct costs. The problem arose because of the complexity of two sections of the price regulations in title 10 of the *Code of Federal Regulations*. First, section 212.83(e) prohibited the banking of nonproduct cost increases. Second, section 212.83(d) stated that a refiner must charge "base price" (for example, maximum lawful selling price) before any nonproduct cost increases could be passed through in price increases. Finally, section 212.83(d) also set forth a ratio designed to limit nonproduct cost increases for each product category. Some firms applied the ratio, designed to limit nonproduct cost recoveries on particular product categories, to the total cost increase recouped on each product category, thereby prorating recouped costs between product and nonproduct cost increases. The RARP compliance personnel also misinter-

---

[17] Ibid.

[18] See 10 C.F.R., section 212.83.

preted this limitation provision. In effect, this permitted firms which did not pass through their total increased costs to bank a portion of their unrecouped nonproduct costs. As a result, refiners included approximately $1.3 billion of invalid costs in banks, an undetermined amount of which was passed through to consumers. This illustrates the lack of a timely detection of violations caused by regulatory complexity; neither the industry nor FEA knew a violation was occurring.

Failure to verify refiners' base prices to resolve all problems arising out of the banking provisions does not preclude the audit of passthroughs of many cost types: imported and domestic crude, semi-finished products and blending items, additives used in the refining process, other purchased costs such as residual fuel and finished gasoline, increases in labor, interest, marketing, utilities expense, and pollution equipment. FEA maintains that it can therefore audit refiners to see that excessive cost increases are not added on to the yet unaudited base price.

Nevertheless, the agency has undertaken to enforce price regulations based in part on a base price which will have taken more than three years to verify. Certain rules for computing that base price, moreover, are still unclear. The banking regulations add another layer of complexity to the enforcement process. These enforcement problems, taken together, cast serious doubt on the clarity of the rule to be obeyed and the certainty of enforcement efforts essential to any compliance effort.

2. Enforcement of the retailer/reseller regulations has been superseded by market forces: The primary enforcement problem with the price regulations downstream from the refiner is that they are generally irrelevant under current marketing conditions. At present, there are adequate supplies of most products at or below the maximum lawful selling price for resellers. Consequently, very few resellers and retailers have kept records adequate to verify their lawful selling price, as required by FEA regulations. These factors have led FEA to cease active enforcement of the reseller/retailer bookkeeping requirements,[19] with the primary enforcement effort for these firms being limited to auditing of resellers to determine compliance during the embargo period.

It should be noted, however, that the current reseller/retailer regulations do contain provisions for cost banks (in a simpler form than that for refiners) and for the determination of classes of purchasers. Although such requirements do not entail as many compliance difficulties as does their application to refiners, continuation of

---

[19] Interview with Gordon Harvey, September 22, 1976.

these rules at the reseller/retailer level will only hinder efficient enforcement of price regulations at this level during a future shortage.

Changed distribution patterns, outdated base periods, inflated allocation entitlements, and inadequate records would prevent effective enforcement of the allocation regulations below the refiner level.

1. FEA product allocation regulations are essentially designed to distribute product equitably among resellers and consumers in a time of shortage but rely on outdated distribution patterns and base periods: FEA product allocation regulations were written during the embargo. Their purpose was that of distributing scarce petroleum products in a manner which took into consideration both the need to establish priorities among various ultimate users and the need to assure the continued competitive viability of the various segments of the petroleum chain of distribution.[20]

The allocation regulations are based on two concepts: the maintenance of relationships between suppliers and purchasers and the allocation of product through the determination of an allocation entitlement. The regulations contemplate requiring suppliers (sellers) to make covered products available during the present time period to the same purchasers (buyers) they sold the product to during a specified base time period. This established the supplier/purchaser relationship.

Once these supplier/purchaser relationships were established— that is, firms knew from whom they could buy and to whom they must sell, if requested to do so—the available supply of the product was allocated. The major components of this allocation concept are (1) allocation entitlements, (2) allocation levels, and (3) allocation fractions.

Allocation entitlements are based on either "base period use" or "current requirements." Base period use is the amount of particular product purchased or otherwise obtained during the base period by a particular purchaser. Base period use must be computed for all purchasers except those entitled to current requirements. The rationale underlying the choice of the currently prescribed base period for each product is that it represents the last pre-embargo time periods of normal business operations. Allocation entitlements of suppliers may

---

[20] Like the crude oil pricing regulations, the crude oil allocation regulations are enforceable. The supplier/purchaser freeze has a current base date, February 1, 1976; domestic (though not imported) crude oil is in short supply; and the number of purchasers is limited. Likewise, the limited number of purchasers and sellers makes the interrefiner buy/sell program workable from an enforcement point of view. The same is true of the crude "entitlements" program to reduce refiner crude cost differences, which is technically an allocation program.

be adjusted because of certification of current requirements by their purchasers.

Although the allocation regulations require suppliers to sell each customer his product allocation on demand, the regulations do not require the customer to purchase his allocated volume. If surplus supplies of product are available, he may purchase them from any willing seller.

Consequently, the passage of time has created two fundamental defects in the present allocation structure. First, for those purchasers not granted their "current requirements" of product, the base period levels are completely out of date, even if base period supplier/purchaser relationships are maintained. The base period for most products is 1972 and thus would be at least five years old in any future supply interruption.

Second, the ability of purchasers to obtain "surplus" product from any available source has permitted many purchasers to obtain product according to price and service, rather than according to their record base period relationships. Even if a purchaser elects to buy product elsewhere, the supplier is still required to reflect base period purchasers on its supplier/purchaser relationship lists and to make product available to these purchasers on demand. Rules for terminating supplier/purchaser relationships require mutual consent of the involved parties and/or FEA approval.

Such consent may not be given even when relationships change in actuality. Since purchasers are not required to buy product, suppliers are sometimes anxious to terminate the relationships and thereby avoid the requirement to retain product for the purchaser. The purchaser, on the other hand, may wish to retain the relationship as a hedge against a future shortage while purchasing product in a period of ample supply from another supplier. Conversely, a purchaser may find a better supplier (lower price, better service, et cetera) who is willing to supply him. However, the existing supplier may wish to retain the customer as a hedge during a shortage, particularly if the supplier has a relatively high maximum permissible selling price.

Thus, the allocation mechanism is a system which nominally guarantees future supply rights which bear little relationship to current patterns of distribution.[21] Firms are buying from suppliers other than those to whom they are legally tied by the regulations. Similarly, many firms are purchasing volumes significantly in excess of their allocation entitlements based on historical use. Hence, in the event of a future shortage, any attempt to enforce the current allocation

---

[21] Interview with Gorman Smith, October 13, 1976.

regulations as to base period supplier/purchaser relationships and quantities, even if possible, would result in a severe disruption of current supply patterns.

2. Attempts to adjust allocation levels within the current regulatory framework may undercut FEA's ability to allocate product in a shortage: After the embargo, the basic elements of the allocation program were retained although product supplies began returning to normal levels. As a result, many businessmen, desirous of conforming their "official" FEA base period relationships to current practices in order to prevent disruption in the case of another shortage, petitioned FEA for revisions in their supplier/purchaser relationships. Consequently, FEA's emphasis shifted from the embargo function of monitoring complaints and directing actual allocations of products to that of updating allocation entitlements and providing assignments for new businesses to reflect more accurately changes in business operations since the base year of 1972–1973.[22]

Such adjustments, however, have had the effect of undercutting the ability of the current regulatory structure to distribute product in another shortage. In the case of gasoline retailers, new market entries, which had been halted during the actual shortage, began to appear in response to the then ample supply conditions. FEA realized that while the regulations permitted such entries a recurrence of supply disruption would eliminate the surplus. In such a situation, only firms which had allocation entitlements based on their 1972 or 1973 operations would qualify for allocations. New entries would be unable to get product and would fail.

Consequently, regulations were developed to monitor and establish entitlements for new gasoline stations, thereby including them in any allocation system which might be required in the future.[23] Approvals of entitlements for new entrants, however, were granted with little regard to whether their inclusion would lessen the effectiveness of the allocation scheme during a supply shortage.

Additionally, under FEA procedures, a switch in base period relationships is automatically granted when there is agreement among the purchaser, the current supplier, and the prospective supplier. According to regional officials, the procedures for handling three-party agreements vary among regional offices. It should be noted that, where a three-party substitution is made, the party gaining the new customer may certify the needs of that new customer to his own supplier, thus increasing his allocation. On the other hand, the obligation

---

[22] Ibid.

[23] 1 FEA Guidelines, par. 13245, appendix.

to request a reduction in the allocation to the old supplier apparently rests with the old supplier,[24] and he may not request a reduction. In such a case, there is a double counting of the allocation for the switching customer, thus further distorting the allocation system. A dispute apparently exists among the regions as to this practice, with some regions apparently hesitant to permit the double allocation, although the regulations seemingly permit it.

Finally, those purchasers entitled to their current requirements of a product may overstate their needs. It is not surprising, then, that total allocation requirements are overstated: for gasoline, present entitlements are estimated to exceed current usage by approximately 30 percent.[25] Although this overstatement of requirements can be tolerated under ample supply conditions, in a shortage situation, it would severely distort the allocation process and result in artificially low allocation fractions.

3. Today, in many cases, there are insufficient records of base period relationships to allow enforcement of the allocation regulations: Enforcement of the allocation regulations would require the reconstruction of supply relationships which existed four years ago, yet many smaller firms today have no adequate records documenting the 1972–1973 supply relationships. The agency has had the occasion to try to reconstruct base period supply relationships using existing records during routine audits or in response to a complaint, and the necessity for such reconstruction has not, in a given case, precluded investigation of a potential violation. Nevertheless, since FEA does not routinely audit these records,[26] and since they are not a normal part of a firm's business operations, their adequacy for industrywide enforcement, as in a shortage situation, is a matter of speculation.

Commonly used and available sanctions fail to deter violations and encourage voluntary compliance.

1. The practical need for deterrence: In order for a regulatory program that governs primary business activity during a time of crisis to achieve its goals effectively, compliance must of necessity be voluntary. That is, businesses must have an incentive to obey regulations at the time their business activity is carried out, rather than merely to adjust their behavior retroactively in response to a remedial order from the agency. When business activity occurs during an emergency, securing compliance then rather than some later time is

---

[24] 10 C.F.R., section 211.13(f).

[25] Interview with Gorman Smith, October 13, 1976.

[26] Interview with Gerald Emmer, director of allocation regulations, FEA Office of Regulatory Programs.

the only sure way to be certain that congressional and regulatory objectives are being achieved. Consequently, the penalties for being caught violating regulations must be such that businesses find it disadvantageous simply to ignore a rule and await detection and subsequent sanctions.

What little evidence there is appears to demonstrate that the amount of punishment meted out to violators of price restrictions must be especially severe to achieve compliance. A study of price violations during World War II showed that, in districts where enforcement was lax, there were many violations of the price regulations. In contrast, in areas where violators were sent to prison, the rules were obeyed.[27]

The need for such "general deterrence" is particularly important with respect to FEA regulation. During an actual shortage, the agency will likely devote most of its resources to allocating product and monitoring major pricing problems; therefore, the likelihood that an individual violator will be the subject of an audit and remedial order is small. Even after the end of the crisis, not all violators will be detected. Although FEA plans to audit all major refiners to determine whether they violated the price regulations during the past embargo and subsequent periods of control, current FEA work plans call for a volume of audits sufficient to discover only 42 percent of the violations committed by resellers during the embargo. Moreover, these planned audits will not be completed until the end of fiscal year 1978, more than four years after the end of the embargo. If voluntary compliance with FEA regulations is to be achieved, then the sanctions imposed for violation of FEA regulations must be sufficient to deter a businessman from violating them, even if the chance of detection is less than 50 percent and even if the imposition of a sanction may occur considerably in the future. Unfortunately, present FEA sanctions are inadequate to the task.

2. The presently utilized sanctions fail to achieve deterrence: The FEA has statutory authority to seek, in conjunction with the Department of Justice, civil and criminal penalties.[28] Furthermore, the agency may seek injunctive relief,[29] and it has much flexibility to shape its administrative remedies.[30]

---

[27] M. Clinard, *The Black Market* (1953), pp. 59-60 and 253-54, cited in William J. Chambliss, "Types of Deviance and the Effectiveness of Legal Sanctions," *Wisconsin Law Review*, 1967, pp. 703 and 710.

[28] EPAA, section 5(a)(3).

[29] 10 C.F.R., section 205.204.

[30] 10 C.F.R., sections 205.191-197.

In spite of the need for substantial penalties to deter violation of the petroleum regulations, and in spite of the above broad enforcement powers, the sanctions actually invoked by FEA often may result in little or no financial impact on the firms involved. With respect to refiners, FEA has permitted a retroactive regulation change and class exception that nullified violations discovered by compliance personnel or softened the regulations. In this instance, the regulations were amended to allow refiners to bank nonproduct costs, beginning February 1, 1976, and it is proposed that exceptions be individually granted to permit refiners to bank unrecouped nonproduct cost increases from an even earlier date.

Even where a violation proceeding is carried to its conclusion, the remedy ordered may take effect only on paper. For example, adjustment to banks is a commonly ordered remedy. FEA reports that from July 1, 1974, to September 30, 1976, it had ordered adjustments totaling $459.1 million;[31] of those adjustments, $306.2 million, or 67 percent, were bank adjustments. In a period when market forces are holding many product price levels at or below the regulatory ceiling, adjustments to banks have no present actual effect on the firm. Moreover, this remedy is not reserved for violations involving miscalculation or fraudulent alterations in banks; it may be ordered, for example, to remedy a price regulation violation involving actual cost benefits to the firm. It is doubtful, then, that the perceived threat of this sanction will effectively deter potential violators.

The next two most common remedies are price rollbacks and refunds, amounting to $100.3 million and $52.6 million, respectively, between July 1, 1974, and September 30, 1976. Imposition of these sanctions, while it does involve actual financial impact on the firm's future receipts and disbursements, puts the firm in no worse position than it would have been had it complied with the regulations. And, to the extent that interest charged for the duration of the violation is less than the time value of the overcharge to the firm, the firm is benefited.

Penalties collected do involve an actual loss to the violator, and FEA claims seek the imposition of such penalties where appropriate. For the time period discussed above, however, FEA reports penalties of only $2.9 million, of which $1.5 million was imposed upon producers and major refiners.

Enforcement Problems Created by the Present Regulatory Structure Will Impair an FEA Compliance Effort in a Future Shortage. The task

---

[31] FEA, Summary of Compliance Effort, provided to the task force.

force concludes that regardless of the success of FEA's compliance work plan in achieving present audit goals, various aspects of the product price and allocation regulations make their present enforcement difficult. More importantly, these problems will severely undercut FEA's ability to achieve an effective, rational energy program in the event of a future petroleum shortage.

The complexities introduced into the price rules, particularly at the refiner level, by the class-of-purchaser rules (and delays in defining its underlying terms) and by the provisions governing cost banks, have precluded verification at the present time of the May 15, 1973, base price for refiners and delayed verification of cost passthroughs from banks.

Enforcement of the refiner maximum lawful selling price rule, then, cannot begin until at least January 1, 1977, the scheduled audit completion date for the May 15 price. Even at that time, effective enforcement will not be automatic. Litigation over the complicated rules for computing the maximum price may push final remedial orders back many months.

On top of these problems, the refiner cost banks add another layer of complexity. As discussed above, the interrelationships among banks and the effects of product decontrol on the banks magnify the potential for good faith disputes over enforcement.

In a future shortage, FEA could take the refiner base price as given, regardless of inaccuracies, and enforce price levels downstream from the refinery by auditing cost passthroughs. Enforcement at these lower levels may be aided by the high visibility of violations and the absence of problems in computing refinery base prices. The problems with banks, however, where used, persist at these levels. Furthermore, enforcement difficulties increase downstream as the number of firms increases. Attempts to maintain, during a time of normal supply, a system of product allocation based on outdated base periods have resulted in distortion of the allocation scheme. Moreover, the allocation system, even as adjusted, no longer reflects current distribution patterns. Indeed, many small businesses may not have records adequate to establish base period relationships.

Consequently, the present regulatory structure would not provide an adequate foundation upon which to build a compliance program in a future shortage: the agency could not presume that current prices were lawful, nor could it expect to allocate product according to "record" relationships without great disruption to current distribution patterns.

106

Just as importantly, FEA lacks a system of meaningful sanctions. In order for businessmen to have an incentive to comply voluntarily with a regulatory scheme, the expected costs of violating a rule must outweigh the benefits of so doing. The major FEA sanctions for price violations (rollbacks, refunds, and bank adjustments) merely return the violator to the status quo—he is no worse off than if he had obeyed the rule. Thus, if the violator believes he has a reasonable chance of escaping a remedial order, it is in his interest (apart from adverse publicity) to violate the regulation. The complexity of current FEA price regulations only serves to enhance the potential violator's incentive to violate a rule. FEA determination of a violation is thought to come long after the fact and, if the violator is caught, the regulations' ambiguities give him the opportunity to challenge the agency's allegations.

In sum, unless FEA's regulations are simplified, base periods are made current, and more effective sanctions are established, FEA's enforcement effort, no matter how well structured and staffed, will face great difficulty in assuring compliance with the regulations, both today and in a future shortage.

The Exception Process

Introduction. Under the Federal Energy Administration Act of 1974, FEA is required to grant exception relief for those unfairly affected by FEA regulations:

> (D) Any officer or agency authorized to issue the rules, regulations, or orders described in paragraph (A) shall provide for the making of such adjustments, consistent with the other purposes of this Act, as may be necessary to prevent special hardship, inequity, or unfair distribution of burdens and shall, by rule, establish procedures which are available to any person for the purpose of seeking an interpretation, modification, recision of, exception to, or exemption from, such rules, regulations, and orders.[32]

In fulfilling its obligations under this congressional requirement, the Office of Exceptions and Appeals has efficiently handled the applications of individuals who feel unjustly affected by FEA regulations. In general, the office has established a target for passing on applications for exception relief within sixty days after the application has

---

[32] FEAA, section 7(i)(1)(D).

been perfected.[33] Information supplied to the task force by the Office of Exceptions and Appeals indicates that over the past year, the average time for processing cases has been about fifty days.

The exception process, like other FEA programs, has its roots in the Economic Stabilization Act of 1970. Section 203 of that act, which authorized the President to stabilize prices, rents, wages, and interest rates, required that the President provide for the making of exceptions to the program to prevent "gross inequities [and] hardships." [34]

The implementation of this provision under the original wage and price freeze resulted in very few grants of exceptional relief. A stringent interpretation of the relief provision was consistent with the 1971 view of the control program; at the outset, controls were to be imposed for but a ninety-day period. The Cost of Living Council, exercising the authority delegated to it by the President under the statute, authorized exceptional relief only in cases of "serious hardship" or "gross inequity." As implemented by the CLC under the price freeze program, then, these narrow standards contributed to an application of the freeze which was simple and universal, with little allowance for exceptional individual circumstance.

As price controls continued beyond the initial freeze, the hardships resulting from long-term universal application created substantial pressure for adjustments to meet individual circumstances not contemplated by the emergency regulatory scheme. These adjustments were made on a general basis through changes in the price regulations themselves and on a case-by-case basis by means of the Price Commission and CLC exception processes.

When comprehensive price and allocation regulations for the petroleum industry were adopted by the new FEO in January 1974, an exception process was established by rule. The FEO Office of Exceptions and Appeals adopted the standards for granting exceptional relief which were at least nominally the same as had been the practice under the Economic Stabilization Program; namely, the criteria of "serious hardship" and "gross inequity." [35] These tests initially performed the function of limiting relief to those parties who to an unusual degree were harmed by the program or suffered consequences not intended by the drafters of the regulations.

---

[33] Of course, many applications are received without sufficient information to enable the office to rule on the requests. As a result, there may be a delay between the time the applicant first submits his request and the point at which FEA considers it has sufficient information to permit processing of the application.

[34] P.L. 92-210 (1970).

[35] FEA has occasionally also granted relief to alleviate an "unfair burden." See FEA, Office of Exceptions and Appeals Guidelines, 41 Fed. Reg. 40858 (1976).

"Serious hardship" relief required a showing that the operation of the regulations would cause a particular party to suffer serious financial difficulty, often operation at a loss and possible bankruptcy. Such relief was a necessary adjunct of a program which established arbitrary base period relationships for the pricing and the allocation of petroleum. Under such programs, it was expected that many, if not most, individuals would experience some hardship, particularly in a time of shortage. On the other hand, relief is appropriately granted to those individuals who find that the operation of the regulatory program will cause them hardship severe enough to threaten their financial viability.

Additionally, the operation of generalized rules in particular factual circumstances may produce results which were unintended or undesired by the writers of the regulations. In a rapidly changing shortage situation, conditions also may become such that previously appropriate rules are no longer suitable guides to business behavior. As the conference report on ECPA noted: "the exceptions process is designed in substantial measure to resolve factual situations which could not have been and were not contemplated at the time the general statutory or regulatory programs were adopted." [36] Early exceptions cases held that such "gross inequity" relief required a greater showing than that a firm was affected in a manner different from that of other firms. A comparative test was to be used:

> The application of any general regulatory provision whatsoever will affect different firms in a different manner depending on the nature of their operations. The effect of any such regulation will also result in a particular firm being more adversely affected than any other firm. This result is an unavoidable element of the regulatory process, and exception relief is not appropriate under these circumstances solely on the basis of a showing that on a scale which measures the adverse impact of a particular regulatory provision a given firm is more adversely affected than other firms.
>
> . . . Rather, a firm must demonstrate that the application to it of a specific regulation significantly and uniquely impedes its operations, or results in a serious competitive disadvantage to the firm in relation to other similarly situated firms, or seriously frustrates the effectuation of the objectives which the regulation was designed to achieve.[37]

---

[36] H.R. Rep. No. 94-1392, 94th Congress, 2d session (1976), p. 60.
[37] Quaker State Oil Refining Corp., 2 FEA, par. 83,808 (August 18, 1975).

In this light, a finding of serious hardship can be viewed as a special case of gross inequity relief: since it is not intended that the operation of the regulations, as such, endanger the financial viability of individual firms, those firms suffering a serious financial hardship are, in effect, presumed to be treated in a grossly inequitable manner by the regulations without a showing beyond the mere fact of adverse financial impact.

The Changing Role of the Exception Process. As FEA has recently noted:

> The exceptions process is continually evolving. . . . Through the method of case-by-case determination, the FEA has attempted to balance the intended goals of the particular regulations from which relief is sought, the need for regulatory uniformity, the specific policy goals of the Agency, the current and projected condition of the particular markets affected by the individual application, and the impact of the Agency's action on the application, other members of the petroleum industry, and consumers.[38]

The exception process has thus reflected the varying roles played by FEA during its existence: from a crisis manager during the embargo to an administrator of longer-term national energy policy at present.

The initial regulations promulgated by FEO in January 1974, during the embargo, were generally viewed as emergency rules of relatively short duration. During this period, the exception process played a key role in adapting to particular situations universal rules which were of necessity arbitrary and sometimes hastily drafted.

Following the end of the embargo, it became increasingly apparent that much of the regulatory framework was to remain in place for a considerable period. Consequently, the agency needed to adapt short-term regulations to the needs of a long-range regulatory program. In part, this function was carried out by FEA's regulatory development process. In part, it was also accomplished through the exception mechanism. The regulatory development process was often haphazard and slow to respond to changing market conditions and to the requirements of a long-term regulatory system. As a result, there were increasing demands for exceptional relief from those adversely affected by the regulations.

We believe it can be fairly stated that the Office of Exceptions and Appeals currently views its function as not only to provide

---

[38] FEA, Office of Exceptions and Appeals Guidelines, 41 Fed. Reg. 50856, 50862 (1976).

"fairness" to individual firms injured by regulation but also to serve as a "cutting edge" for the development of new regulatory policy in circumstances where the generalized impact of a particular rule is not known. The opportunity for an evolutionary trial-and-error process is seen as an advantage of adjudicatory rulemaking, as it has the effect of allowing the codified regulatory structure to remain unencumbered by a large number of provisos and alternative rules.

Although the making of policy on a case-by-case basis, as opposed to a formal rulemaking, may have merit in some circumstances, the task force believes that the general evolution of the exceptions process from one which grants truly "exceptional" relief toward one which develops rules of general application to large numbers of individuals has given rise to serious problems.

Currently, the Exception Process Develops Rules of General Application Rather than Merely Granting Relief in "Exceptional" Circumstances. A review of exceptions decisions clearly shows that the exception process has become a supplementary method of establishing rules of general or potentially widespread application. That is, the exception process has established various types of relief which FEA expects to grant on a relatively automatic formula basis without requiring an exception applicant to show anything more than that he comes within the formula.[39]

We do not claim that it is inherently inappropriate to use the exception process in this manner.[40] We do, however, believe there are some circumstances in which regulatory development on a case-by-case basis limits full prior consideration of alternative regulatory policies and measures of relief and gives rise to other significant problems. We thus urge that the implications of such use be clearly understood and that FEA decide formally the extent to which it wishes to utilize the exception mechanism as an integral part of the rule-making process.

Four lines of cases exemplify the types of "rulemaking" currently accomplished by the exception process: (1) the granting of relief to small refiners who are net purchasers of crude oil entitle-

---

[39] See, for example, the general categories of relief in FEA, Office of Exceptions and Appeals Guidelines, 41 Fed. Reg. 50858 (1976).

[40] Many federal regulatory agencies, such as the Interstate Commerce Commission, National Labor Relations Board, and Civil Aeronautics Board, utilize an adjudicatory process as a useful supplement to formal rule-making procedures. Indeed, the Supreme Court, while noting a general preference for the rulemaking process as a means of establishing regulations of general application, has recently reaffirmed the legality of agency rulemaking by adjudication. NLRB v. Bell Aero-Space Co., 416 U.S. 267, 293 (1974).

ments, a decision to subsidize this group; (2) providing relief designed to promote business activity which will achieve national energy goals even though the applicant, himself, is not hurt by the operation of the regulations, an implicit decision to introduce rate-of-return regulation; (3) the modification of refinery cost rules to permit allocation of costs on the basis of refinery yields, a correction of an FEA oversight; and (4) the use of the exception process to provide relief to operators of natural gas liquids (NGL) plants when their costs exceed a fixed ceiling on price increases, the utilization of the exception process to make an automatic adjustment to price ceilings.[41]

Excepting small refiners from the obligation to purchase entitlements: a subsidy program. A very significant instance of exception "rulemaking," in terms of economic impact, is relief for small refiners from entitlements purchases when purchases would result in the refiner's achieving less than its historic profit level. The leading case is *Delta Refining Co.,*[42] establishing the rule that:

> Exception relief from the purchase requirement of the entitlement program will generally be granted to small refiners if the application of the program's requirements would otherwise prevent the firm from achieving the lesser of its historic profit margin or its historic return on invested capital.[43]

This standard, as amplified, has been followed in a large number of cases [44] to provide exception relief sufficient to allow a small refiner the lesser of its historic profit margin or the average return on invested capital in the best four of the most recent seven years.

---

[41] While the exception process has often developed rules through the adjudicatory process, it has on several occasions operated to bring to the agency's attention matters requiring basic regulatory changes. For example, see New England Petroleum Co., 2 FEA, par. 83,136 (May 2, 1975), discussing the problems of competitive imbalances in the East Coast residual fuel oil market, and 41 Fed. Reg. 30321 (July 23, 1976), amending the entitlements regulation to alleviate the problem; Iowa Southern Utilities Co., 2 FEA, par. 83,187 (June 16, 1975), granting relief from the low-sulfur petroleum products regulation, and 41 Fed Reg. 24517 (1975), rescinding that regulation.

[42] 2 FEA, par. 83,275 (September 11, 1975).

[43] 2 FEA, pars. 83,883-84 (September 11, 1975).

[44] For example, Farmariss Oil Corp., 2 FEA, par. 83,277 (September 18, 1975); Fletcher Oil and Refining Co., 2 FEA, par. 83,278 (September 16, 1975); Laketon Asphalt Refining, Inc., 2 FEA, par. 83,281 (September 17, 1975); Midland Cooperatives, Inc., 2 FEA, par. 83,283 (September 17, 1975); Mohawk Petroleum Corp., 2 FEA, par. 83,284 (September 16, 1975); Pasco, Inc., 2 FEA, par. 83,290 (September 17, 1975).

Additionally, FEA has extended the rule to new small refiners having no profit history. In *Wickett Refining Co.*,[45] FEA granted a small refiner, which had recently entered the market, relief from its projected entitlements obligation to the extent necessary to permit it to achieve the average return on invested capital realized by other small refiners, an amount which the agency computed is approximately 10.9 percent.

FEA recognized the rulemaking nature of the standard announced in the *Delta* case. The Office of Exceptions and Appeals published a *Federal Register* notice requesting comments on the standards to be used in granting relief from entitlements purchases by small refiners.[46]

Through this holding, the exception process has in essence amended the entitlements program to establish a program for assisting small refiners in maintaining their historic profit positions. These exceptions are granted on a periodically renewable basis and, most significantly, there is no required showing that the operation of the FEA regulatory program, itself, has caused the lack of funds necessary to permit purchases of entitlements.

Allowing the sale of crude oil at greater prices to encourage additional investment in crude production: establishing rate-of-return regulation. FEA has established the principle of granting relief from lower tier crude oil pricing restrictions where significantly increased production costs leave a firm with little or no economic incentive to produce crude oil from existing wells on a developed property.[47]

---

[45] 2 FEA, par. 83,238 (August 6, 1975).

[46] 40 Fed. Reg. 33489 (1975).

[47] The relief granted is in the form of authorization to sell some or all of the lower tier crude oil at upper tier prices. FEA, in effect, has been requested to extend the principal so as to grant exception relief from the upper tier ceiling price provisions as well; in Bruce A. Clark, 4 FEA, par. 83,033 (August 6, 1976), no indication was given that such an extension would not be made in an appropriate case, although relief was denied on the ground that an actual disincentive to produce had not been shown.

Leading cases involving production disincentives due to increased operating costs are Pruet & Hughes Co., 2 FEA, par. 83,270 (August 25, 1975); and Braden-Deem, Inc., 3 FEA, par. 83,072 (January 15, 1976). These are frequently cited in the context of the statement that "the regulations which FEA has promulgated are clearly not designed to result in a reduction in the Nation's supply of recoverable domestic crude oil," as establishing that exception relief will be granted: "where a strong and convincing showing is made that: (i) under FEA crude oil pricing regulations a firm has little economic incentive to continue to produce crude oil; (ii) there is little possibility that the crude oil in the field could be recovered except through the continuation of the firm's extraction operation; and (iii) the wells involved are already part of a continuing extraction operation." See, for example, William H. Player and Associates, 3 FEA, par. 83,164 (April 9, 1976).

Recently, also through the exception process, FEA has extended this rule to permit producers wishing to *expand* production on existing properties to sell lower tier oil at upper tier prices in an amount sufficient to permit an adequate return on the required investment.

This extension was begun in *A & N Producing Services, Inc.*[48] There, a lease which had been in operation for a number of years was shut down because of equipment difficulties. A showing was made that the only feasible way to recover the crude oil remaining in the reservoir was by drilling a new well to replace the existing well. FEA found that a return on the additional capital required to drill the replacement well was necessary in order to prevent that lease from being abandoned. Consequently, it permitted the operator of the property to sell at upper tier prices an amount of lower tier oil sufficient to achieve a 15 percent return on the investment.

Then in *Austral Oil Company, Inc.*,[49] FEA granted a producer the right to sell otherwise old oil at upper tier prices in amounts sufficient to enable him to install additional water-flooding projects on his property and thereby to *increase* the volume of oil produced. As in the *A & N* case, FEA noted that their relief would generate revenue sufficient to permit an average return on invested capital of "approximately 15 percent." It was also noted that "in view of the risk involved in the venture and other alternate investment opportunities which Austral might consider, such a rate of return would not produce windfall profits in this case."[50]

The *Austral* case establishes a precedent which is a sharp departure from prior exception cases.[51] In previous cases, the purpose of the relief was to prevent the operation of the price ceiling from causing a producer to abandon his *existing* investment and cease producing current volumes of crude oil. Such a circumstance is more easily seen as a "hardship" or an inequitable result for such producers, since for other producers the lower tier price levels are sufficient to permit them to remain in business indefinitely. In *Austral*, however, the agency assumed the applicant had alternative investment opportunities for its funds. The operation of the price ceiling merely affected the relative profitabilities of various investment opportunities. Had the exception not been granted, the applicant would have invested his money elsewhere at what the decision implicitly assumes would be

---

[48] 3 FEA, par. 83,172 (April 26, 1976).

[49] 4 FEA, par. 83,004 (July 15, 1976).

[50] Ibid., par. 83,018.

[51] The precedent has been followed in Central Crude Oil Co., 4 FEA, par. 83,104 (September 21, 1976).

114

an equivalent rate of return. Austral was thus not suffering a hardship.

The decision thus represents a substantive determination by FEA to modify the crude oil pricing mechanism for a single firm solely to encourage additional production from existing properties and thus to further the national energy goal of encouraging greater crude production. The procedure necessarily involves regulation of the rate of return on new producing investments.

Modifying the rules governing the allocation by refiners of increased costs: correcting an FEA error. As in effect in 1975, section 212.83(c) of the FEA price regulations required that refiners allocate increased product and nonproduct costs (insofar as permitted by the price regulations) on the basis of the volume of products sold by the refiner. This rule created problems for several smaller refiners who also purchased products and then resold them. Since the rule spoke of products "sold" rather than "refined," a refiner which purchased large amounts of a product (for example, gasoline) would have to allocate a relatively larger portion of its increased crude costs to gasoline because gasoline would represent a higher proportion of its sales than what would be the case if the refiner sold only those products which it refined. If market conditions prevented a refiner from raising its gasoline prices sufficiently to recover the crude cost incurred in refining (for example, middle distillate) it might experience financial difficulty from the operation of the rule. One such refiner, Pasco, requested an interpretation from the Office of General Counsel which would enable it to allocate increased costs on the basis of refinery yields rather than product sales. The General Counsel denied the requested interpretation and Pasco appealed. The Office of Exceptions and Appeals affirmed the denial on July 3, 1975.[52] The appeals decision was quite specific:

> Any contention that the phrase "total volume . . . sold" should be read in a way which conflicts with its obvious meaning in a manner suggested by Pasco is unjustified. When FEA intended that only refined product be included in the cost allocation factor, it did so explicitly in the case of propane. Adoption of Pasco's construction would extend the exception limited to propane to all other covered products and thus, in effect, rewrite the entire rule. This result is in variance with regulations and may only be accomplished by a rulemaking proceeding.[53]

[52] Pasco, Inc., 2 FEA, par. 80,628 (July 3, 1975).
[53] Ibid., par. 80,935.

Shortly after this ruling, another small refiner submitted a request for an exception so that, as Pasco had unsuccessfully requested, it could allocate increased costs on the basis of refinery yields rather than sales. The exception was granted to the applicant, Vickers Petroleum, both prospectively and retroactively on the grounds that failure to do so would cause the firm to move from a profit to a loss position and that serious hardship thus had been shown.[54]

On February 1, 1976, FEA amended the refiner price regulations to provide, specifically, that refiners could allocate increased crude costs on the basis of either refinery yields or product sales, whichever was more advantageous to the refiner.[55]

In May 1976, FEA decided an exception case involving a request that the applicant be permitted to allocate both increased crude and nonproduct costs on the basis of yields.[56] In its decision, FEA noted that the rulemaking, decided subsequent to the filing of the exception application, had eliminated the need for prospective exception relief with respect to the increased crude costs. The decision went on to find that, absent nonproduct cost relief, the firm would suffer a loss, since market conditions were holding down the price of gasoline, while costs allocated to other products could be passed through in the form of higher prices. The agency granted relief, noting:

> In the absence of prospective exception relief, Crystal would have to forego approximately $xxxxx of net income in the current fiscal year. In view of the close analogy between this case and the *Vickers* decision, as well as the substantial financial impact on the firm and denial of its request, we have determined that Crystal is experiencing a gross inequity justifying the approval of prospective exception relief. The firm should, therefore, be permitted to allocate its increased nonproduct refining costs on the basis of its refining yield.[57]

Thus, FEA found a "gross inequity" where the operation of the cost allocation rule set out in the regulations caused more than a minimal loss of revenue. This result is closely analogous to the crude cost regulation adopted by FEA in the rulemaking process, since that rule permitted a cost allocation on the basis of yields as an alternative to one based on sales. These cases, then, demonstrate

---

[54] Vickers Petroleum Corp., 2 FEA, par. 83,351 (December 7, 1975).

[55] 41 Fed. Reg. 5111 (1976).

[56] Crystal Oil Company, 3 FEA, par. 83,206 (May 24, 1976).

[57] Ibid., par. 83,815. It is FEA's practice to delete actual dollar amounts from its written decisions in order to protect the interests of the firms involved.

FEA's use of both the exception process and rulemaking to effect a substantive correction to a poorly drafted regulation.

Lifting the limitation on nonproduct-cost price passthroughs for NGL plants: using the exception process to make automatic adjustments. Section 212.165 of the FEA price regulations places a ceiling of 0.5 cents per gallon on the amount of increased nonproduct costs which a natural gas plant processor may pass through in the form of increased prices for its natural gas liquids. In the preamble language explaining the rule, FEA stated:

> Any firms that have increased nonproduct costs of gas processing that would justify a greater price increase than 0.5 cents per gallon may request permission to charge higher prices, on a case by case basis through the exceptions process.[58]

In a case brought pursuant to the invitation contained in that regulatory preamble, it was held that FEA

> will, as a general rule, grant exception relief to any gas processing plant which can demonstrate that its nonproduct costs since May 1973 have increased substantially in excess of the \$.005 per gallon passthrough permitted under provision of section 212.165.[59]

Additionally, the decision determined that costs would be calculated on a quarterly basis. Relief was granted for a quarterly period to permit a full passthrough of increased nonproduct costs above the ceiling, except where the excess of increased costs above the ceiling was minimal.

Hence, in this case, the exception process was specifically utilized by FEA to provide relatively automatic relief under an exception rule of general application for a large number of gas plant operators.[60]

The Use of the Exception Process To Make Rules of General Applicability May Lead to Vested Interests in the Regulatory System on the Part of Certain Firms and May, by Incremental Decisions, Change the Nature of FEA Regulatory Policy without a Conscious Agency Decision To Do So. *Rulemaking by the exception process may create vested financial interests in the regulatory status quo.* The principal danger of adjudicatory rulemaking in a program intended to expire is

---

[58] 39 Fed. Reg. 44407 (1974).

[59] Superior Oil Co., 2 FEA, par. 83,865 (August 29, 1975).

[60] See also cases dealing with the passthrough on aviation fuel for fixed-base operators, for example, Butler Aviation Int'l. Inc., 3 FEA, par. 83,044 (December 15, 1975).

that the process may create vested interests (that is, subsidies) for certain segments of the petroleum industry, tending to perpetuate certain types of exception relief and the regulatory program as a whole. The clearest example of this phenomenon concerns the establishment of the *Delta Refining* rule excepting small refiners from entitlements purchases to the extent necessary to maintain historic levels of profitability. A similar subsidy may result from decisions assigning resellers to lower cost suppliers in order to maintain the reseller's profit levels.

As this report makes clear, the bias in favor of small refiners already existing in the entitlements program is greater than necessary to maintain a competitively viable small refining sector. When faced with applications by small refiners who found that payments for entitlements would reduce their profits below historic operating levels, the agency made the decision to relieve them of purchase obligations to the extent necessary to allow them to retain their historic levels of profitability.

It is clear that the *Delta* rule is now a continuing subsidy. Since money is fungible, one might conclude from a firm's inability to achieve its historic profit levels if it had to pay entitlements that it is the firm's tax liabilities—not FEA regulations—that impede its ability to earn the customary level of profits. Indeed, the purpose of the crude entitlements program is to minimize crude cost differentials among refiners, a situation created by crude price controls. Small refiners are additionally assisted by means of the "bias" in their favor. If a small refiner experiences reduced profits as a result of this system of cost equalization, it is likely inefficient or poorly positioned in the marketplace.

The nature of the exception process, itself, may well have made it likely that entitlements relief would become a continuing subsidy. The FEA procedural regulations require exceptions applications to be decided consistently with prior cases.[61] Once having decided to allow exception relief to one applicant on the basis of historic profitability, the exception process allows little freedom for the agency to change its decision without a detailed showing of why it was inappropriate to have granted such relief in the first place. The only question apparently remaining after the initial determination to provide relief was the narrow one of deciding on the appropriate measure of "historic profitability." [62]

---

[61] 10 C.F.R., section 306.66(b)(3).

[62] H.R. Rep. No. 94-340, 94th Congress, 1st session (1975), p. 47.

Just as importantly, once the policy of granting such relief was established, the small refiners dependent on the initial decision for the maintenance of their historical rates of profitability developed a strong interest in the continuation of the policy. Likewise, other refiners, whose operating results were perhaps not as severely affected by entitlements purchases as the initial applicants, but whose profits had been reduced somewhat below historic levels, had an incentive to request that relief be extended to them. As soon as this was done, a whole class of small refiners had a strong interest in seeing the operation of this rule continued.

In passing EPCA in December 1975, Congress modified the entitlements program to alleviate what it felt were the unexpectedly harsh effects of the entitlements program on small refiners. Section 403 of EPCA exempted from any entitlement purchases refiners having less than 100,000 barrels per day of output. Recognizing that the specific figure was hurriedly arrived at, Congress provided in section 455 of EPCA that FEA could modify this figure subject to congressional veto. Modifications were recommended in Energy Action No. 2 in May 1976. These amendments, which were allowed to go into effect by Congress, eliminated the blanket exemption and, instead, increased the amount of extra entitlements issued to small refiners under the bias program.

In submitting the bias proposal to Congress, FEA noted the continuation of the exception process, and some senators noted that small refiners could, in the future, obtain exception relief if that were justified. Since the adoption of Energy Action No. 2, FEA has continued to grant relief from the purchase of entitlements under the *Delta* rule.

Thus, the utilization of the exception process to alleviate hardships for some refiners appears to have led inexorably to a general policy of relieving small refiners from purchase obligations to the extent necessary to insure their historic profitability. Such relief is granted as a matter of course, regardless of special circumstances, and despite the fact that the FEA has already devised, as part of the entitlements program, a bias in favor of small refiners to improve their competitive position (a bias which, this report has noted, is already too large in amount).

Another example of an exception rule which may lead to the creation of a continuing subsidy is the assignment of a lower cost supplier when the price charged an independent marketer by its base period supplier is so high, relative to prices charged to its competitors, that the marketer cannot effectively compete in its area of business. This practice had its beginning prior to the establishment of the cost-equalizing entitlements program. For example, in *A-1 Oil Corp.*,[63] the applicant stated that its supplier was obtaining gasoline refined solely from imported crude oil, rather than from a mix of low-priced domestic and higher-priced foreign crude. Consequently, A-1 was at a sharp competitive disadvantage compared with other marketers who purchased gasoline in part made from controlled crude oil. FEA found that A-1 was suffering a serious hardship and ordered the allocation of 50,000 gallons per month of gasoline from suppliers whose prices were within the range of those charged to the firm's competitors.

This type of relief, however, continued after the entitlements program had been adopted for the purpose of equalizing crude inputs to refiners. For example, in *Oskey Gasoline and Oil Co.*,[64] the applicant had difficulty meeting the credit terms required by some of its suppliers and stated that these suppliers charged Oskey prices for gasoline and heating oil which were in excess of the prices which Oskey's competitors charged for resale of those products. FEA ordered that Oskey be assigned new base period suppliers whose price for gasoline and heating oil was below that being charged by Oskey's principal competitors, but granted only an amount of relief sufficient to "permit Oskey to realize a gross profit on sales of gasoline and heating oil . . . consistent with its historic levels and . . . to eliminate serious hardships it would otherwise incur." [65]

Likewise, in *Colonial Oil Company*,[66] the applicant, a large distributor of product, stated that, as a result of a switch in suppliers just prior to the establishment of the mandatory allocation program, it had to pay costs which were out of line with the historic differential between its costs and those charged to its competitors. Consequently, it was forced to decrease its retail margins and to experience decreased profits in order to remain competitive, a situation which FEA believed could jeopardize its continued existence as an independent marketer. FEA determined that an amount of product from a lower cost supplier should be allocated to Colonial according to a formula which would permit the firm to operate at a "level reasonably comparable to its historic position." [67] Subsequently, exception relief was renewed and FEA's assistance in maintaining Colonial's profit level continued until

---

[63] 1 FEA, par. 20,604 (May 31, 1974).

[64] 2 FEA, par. 83,114 (April 9, 1975).

[65] Ibid., par. 83,360.

[66] 2 FEA, par. 83,201 (July 3, 1975).

[67] Ibid., par. 83,658.

October 1976. Indeed, FEA continued to grant this type of relief at the end of 1976.[68]

Thus, a procedure which had its genesis at a time when the regulatory program, itself, could be charged with having created price disparities among competing oil firms [69] was continued into a period in which wholesale relationships were essentially governed by the marketplace and not by the FEA regulatory structure. Of course, in an environment in which the agency feels marketplace forces are satisfactory,[70] a regulatory policy that supplies input to a business at a price lower than the price it could obtain in the marketplace is simply a subsidy.

Again, it appears that, once a decision is made to provide profit or margin protection for a given businessman, it is difficult to deny such exception relief to other similarly situated businessmen, who then have an increasing stake in the maintenance of the regulatory structure.

Use of the exception process to make rules through adjudication may commit FEA to making changes in its regulatory program without adequate analysis. In *Austral Refining Co.*, FEA found that the price ceiling on old oil discouraged new investment in previously producing properties. In response to the application before it, the agency decided to grant an exception which would permit Austral to charge a price for production (blending upper tier and lower tier prices) which would enable it to earn a 15 percent return on the new investment.[71] The logic of this decision, coupled with the precedential nature of the exception process, may well result in FEA's establishing a rate-of-return regulation in other sectors of the industry. Such regulation generally involves the determination of "just" or "reasonable" prices which may be charged by individual companies. This determination, in turn, involves calculations of a level of return sufficient to insure profitability and thereby to attract capital.

As the task force has found, many provisions of the FEA regulatory structure discourage investment, such as for desulfurization

---

[68] For example, Jones and Murtha Distributing Co., Inc., 4 FEA, par. 83,076 (August 31, 1976). In that case, relief running at least through November 1976 was granted in an amount sufficient to achieve J & M's historic margins, rather than historic profitability. See Greenville Automatic Gas Company, 2 FEA, par. 83,337 (October 24, 1975); Saveway Gas and Appliance, Inc., 3 FEA, par. 83,150 (March 31, 1976).

[69] That is, lack of entitlements to correct for different crude costs; price controls being a restraint on produce prices, with little surplus product available.

[70] FEA has proposed plans for the decontrol of motor gasoline, 41 Fed. Reg. 51832 (November 24, 1976); the other major products have been decontrolled.

[71] See also Central Crude Oil Co., 4 FEA, par. 83,104 (September 21, 1976).

equipment or the more complicated processes needed to make high-octane, lead-free gasoline. In *Tenneco Oil Co.*,[72] FEA has already established the principle that exception relief from price controls will be provided to permit refiners to operate certain types of equipment so as to promote national energy goals, even though the refiner is not suffering a financial hardship.[73]

There is no logical reason why a refiner wishing to install complex equipment, but finding that under the FEA regulations he has no economic incentive to make that investment, could not request exception relief from price restrictions or entitlement purchase obligations in an amount sufficient to enable it to earn a necessary return to make the new investment worthwhile. Such an extension of the *Tenneco* principle to the encouragement of new investment in the type of refining equipment thought necessary to the achievement of national energy goals would directly parallel the policy extension involved in *Austral*, which dealt with the pricing of crude oil. As previously discussed, prior to *Austral*, FEA had granted price relief for producers who found that they would have no economic incentive to operate existing wells at current rates of production, absent price relief. *Austral* then extended this concept to the encouragement of new production.

It is doubtful that Congress, in enacting the EPAA and its subsequent amendments, intended to establish a system of rate-of-return regulation to encourage investment in various sectors of the petroleum industry. Rather, Congress seems to have relied on the continuation of the system of fixed-tier price levels in the crude area, with limitations on cost passthroughs in other sectors of the industry. Moreover, it set out methods by which FEA could increase the composite crude

---

[72] 3 FEA, par. 83,200 (May 21, 1976).

[73] FEA permitted the applicant to pass through cost increases otherwise prohibited to provide the financial incentive to operate a new hydrodesulfurization unit. The agency found that, in the absence of the exception, Tenneco would not operate these units, but instead would purchase low-sulfur crude oil, the higher cost of which could be passed on to consumers under price regulations. If the cost of operating these units were permitted to be passed through, lower cost high-sulfur crude could be used. As a result, Tenneco might, on balance, be able to lower costs to its consumers, as well as receive some return on its investment in the units. Moreover, FEA found that failure to grant an exception from the passthrough rules would produce: "results which are contrary to important national energy objectives. Tenneco's construction of a hydrodesulfurization unit actually constitutes an improvement in the overall efficiency of the . . . refinery. The firm's utilization of this innovative technology enhances the domestic capacity to utilize high-sulfur crude oil. In addition, since most of the known crude oil reserves in the United States consist of high-sulfur crude oil, the ability of the . . . refinery to use that type of crude oil increases the Nation's self-sufficiency." Ibid., par. 83,786.

price in order to encourage increased production. The decision to commence rate-of-return regulation to encourage new investment changes the nature of petroleum industry regulation, moving it closer toward the regulated utility model. Thus, in making a decision in a small number of cases to encourage additional investment, FEA may have set the basic regulatory program on a new course without a full and complete evaluation of the consequences of such a step.

The Disadvantages of Using the Exception Process for Rulemaking Usually Outweigh Its Advantages. The task force believes that, as shown in the preceding section, the use of the exception process to make rules of general application poses a serious danger for a program which is intended to expire: case-by-case rulemaking tends to create vested interests in the regulatory system and has led to the establishment of regulatory methods suitable for permanent government control of the petroleum industry. Thus, the use of the exception process to make rules of general application can create a regulatory momentum that may prevent the expiration of the FEA regulatory program.[74]

Consequently, FEA should utilize the exception process to make rules of general application only when the advantages of such a practice clearly outweigh its disadvantages. The advantages of case-by-case rulemaking through the exception process have been claimed to consist of the following: first, that it permits FEA to gather information from the applicant's submissions that might be unavailable during the rulemaking; second, that it permits FEA to experiment with the various approaches to the particular problem, whereas in rulemaking relevant experience may be generated only after the rule is adopted; and third, that the use of the exceptions process reduces the regulatory burden on the industry as a whole by decreasing the number of regulations with which the industry must familiarize itself.

We believe that these claims are not wholly supportable. Moreover, the use of the exception procedure has certain detriments: notice of proposed policies to affected persons may not be as satisfactory as would be the case in rulemaking, and the use of the exceptions process may limit the type of remedies available. Additionally, the use of the exception process to effectuate automatic adjustments

---

[74] Additionally, the task force is guided by the belief, adopted by both courts and commentators, that the formulation of rules through the rule-making process is generally preferable to their formulation in an adjudicatory proceeding. See, for example, Frederick Davis, *Administrative Law Text* (St. Paul, Minnesota: West Publishing Co., 1972), section 6.03.

may needlessly complicate filing requirements of those seeking the automatic adjustment.

The utility of the exception process in gathering the information necessary for rulemaking may be outweighed by the limited number of parties participating in an exception case. The existence of a problem created by FEA regulations, arising from either the application of the rule to an unforeseen circumstance or a change in market conditions, may first come to FEA's attention in a request for an exception. When such a request is received, the agency may have insufficient information to formulate a rule of general application. Only after some time, perhaps after evaluating responses to the first decision on the subject, will FEA be in a position to ascertain whether the situation that is complained of involves more than a very few individuals. Additionally, the Office of Exceptions and Appeals has broad power to require that the applicant submit such information as is deemed necessary to decide the case. Consequently, through the use of such requests, FEA may be able to obtain detailed information about the impact of petroleum regulations on the complained-of situation, information that might not readily be volunteered in rulemaking comments (particularly insofar as confidential information is concerned).

Reliance on exception applications, however, may also serve to limit the amount of information available to FEA. Affected parties may fail to submit an exception petition, because they may not realize that an exception is a likely solution to their problem, or because they may be hesitant to submit the detailed financial data that often must accompany an exception petition. The process also tends to exclude information from those who do not have a direct interest in the particular case under consideration, but who for other reasons may have knowledge which would be helpful to the solution of the problem.

Thus, by confining itself to the exception process, even when it appears from applications that a problem of possibly general application exists, the agency may act on less than full knowledge. That the Office of Exceptions and Appeals recognizes this limitation may be seen from the fact that it made a general request for comments concerning the small refiner exception from entitlements purchases before establishing a general rule.[75]

An example which is useful in pointing out the ability of a rulemaking proceeding, as well as the exception process, to gather information involves the problem of nonproduct cost passthroughs by crude oil resellers. As the result of an apparent oversight, crude oil

---

[75] 40 Fed. Reg. 33489 (1975).

resellers are not permitted to pass through increased nonproduct costs.[76] In an application filed in August 1975, one such reseller complained that the failure to permit it to pass through its nonproduct costs placed the firm in financial difficulty, since most of its increased costs were nonproduct costs. FEA decided to grant this firm an exception and permitted it to pass through the total amount of its nonproduct costs as of the date of the application.[77] This case was followed in a similar situation.[78]

Although it might be argued that, because of the limited number of persons involved and the need for further information about the reseller sector of the industry, it was appropriate to develop this type of relief for crude oil resellers through the exception process, rather than undertaking a rulemaking proceeding. Yet, FEA has recently undertaken a general rulemaking inquiry dealing with the impact of price regulation on the resale of crude oil. As part of this inquiry, FEA has undertaken to gather information not only about crude "resellers" but also about crude oil "brokers," a group of individuals about whom FEA apparently did not possess adequate information:

> While not the subject of a specific proposal, FEA also solicits comments and data concerning the operation of firms which are sometimes referred to as "brokers," and which may provide some of the services of crude oil resellers, e.g., transportation of crude oil from the property to delivery point, location of a purchaser, etc., but which do not take title to such crude oil. . . . comments should address the magnitude and nature or the operation of such firms and the manner in which purchasers treat such fees paid in computing the product costs for the purpose of FEA price rules.[79]

Additionally, FEA also endeavored to identify other problem areas about which it had insufficient knowledge:

> FEA also invites comments in this proceeding addressed to any other issues relating to the application of the price regulation to resales of crude oil, even though such issues have not specifically been raised in this Notice. This rulemaking is intended to be comprehensive in nature and FEA requests comments which reflect such a broad scope.[80]

FEA announced that there would be hearings on the topics covered.

---

[76] 10 C.F.R., section 212.93.

[77] Producers Pipeline Co., 3 FEA, par. 83,070 (January 9, 1976).

[78] Marvin E. Boyer Co., Inc., 3 FEA, par. 83,088 (January 30, 1976).

[79] 41 Fed. Reg. 47081 (1976).

[80] Ibid., p. 47082.

The task force believes that the latter approach, that of a preliminary inquiry into a problem area prior to the formulation of specific regulatory language, is an appropriate way of dealing with emerging problem areas about which FEA has not yet compiled sufficient information to develop a remedial rule. Of course, during the pendency of such a rulemaking, FEA could still accept exception applications for interim relief, and the data contained in such proposals, insofar as possible, could be utilized by the agency in the general rulemaking proceeding.

Exception cases show that most of the trial-and-error experience relates to the measure of relief, rather than the decision to grant relief. One of the arguments used by those in FEA who favor rulemaking through adjudication is that the process permits the agency to utilize, in effect, a trial-and-error technique of pretesting regulatory standards prior to the determination of a general rule. The task force believes that this position has merit, but also finds that an adjudicatory approach is more appropriate to formulating technical standards governing the application of a rule—for example, appropriate measuring periods to determine "historic" profit levels—than it is to making major policy determinations. This distinction recognizes that once a decision is made in the exception process to provide a certain type of relief, that determination is difficult to reverse because of the *stare decisis* (standing by, or following, previous decisions) nature of the process. On the other hand, given an FEA decision to provide a certain type of relief, the technical standards to be used in measuring that relief are more susceptible of individual adjustment. Neither the agency nor an applicant has an interest in the utilizing of a formula for relief which does not effectively carry out the policy decision to provide some level of relief in a given situation.

The development of the *Delta Refining* rule governing the exception of small refiners from entitlement purchases illustrates the way in which a general policy may be made relatively rapidly once the existence of a problem becomes known, while formulation of appropriate measures of relief may entail trial-and-error formulation. Small refiner applications for exception from the entitlements program followed almost immediately upon the institution of the program in November 1974.[81] In the early cases, the applicant firm was required to prove that its profit position was adversely affected, however such effect was defined,[82] by its entitlements obligations. When such an

---

[81] Pasco, Inc., 1 FEA, par. 21,668 (December 20, 1974).

[82] The decisions define such adverse effects on small refiners as operating loss, Mohawk Petroleum Corp., 2 FEA, par. 83,049 (February 24, 1975); Famariss Oil

adverse effect was shown, relief would be fashioned in an attempt to allow the firm to achieve its historic financial operating posture as determined by some type of measurement.[83]

In *Delta Refining*, the standards for demonstrating hardship and the method of computing relief merged: the hardship which will entitle a small firm to relief is defined as that which will prevent the firm from achieving its historic return on invested capital; the relief is fashioned to enable the firm to achieve such return.

> Exception relief from the purchase requirement of the entitlements program will generally be granted to small refiners if the application of the program's requirements would otherwise prevent the firm from achieving the lesser of its historic profit margin or its historic return on invested capital.[84]

There have been three significant decision points in the development of the *Delta Refining* rule. First, there was the point at which the agency recognized that the entitlements program, if applied uniformly, would have serious adverse effects upon small refiners whose proportion of old crude oil receipts significantly exceeded the national average. This problem was recognized almost immediately by Special

---

Corp., 2 FEA, par. 83,080 (March 28, 1975); Mid-Tex Refining, 2 FEA, par. 83,090 (March 27, 1975); North American Petroleum Corp., 2 FEA, par. 83,261 (July 31, 1975); breaking even, Beacon Oil Co., 2 FEA, par. 83,060 (March 12, 1975); reduced profits, Husky Oil Co., 2 FEA, par. 83,146 (March 28, 1975); drastically reduced profits, Newhall Refining Co., 2 FEA, par. 83,092 (March 28, 1975); substantially lower profits, Delta Refining Co., 2 FEA, par. 83,078 (March 28, 1975); or reduced profits threatening to impair the firm's competitive position and capital expansion program, OKC Corp., 2 FEA, par. 83,074 (March 21, 1975).

[83] This goal was variously termed historic performance, OKC Corp., 2 FEA, par. 83,074 (March 21, 1975); Delta Refining Co., 2 FEA, par. 83,078 (March 28, 1975); historic financial and operating posture, Beacon Oil Co., 2 FEA, par. 83,060 (March 12, 1975); Young Refining, 2 FEA, par. 83,106 (March 27, 1975); Famariss Oil Corp., 2 FEA, par. 83,080 (March 28, 1975); historic level of profits, Southland Oil Co., 2 FEA, par. 83,099 (March 27, 1975); historic posture and growth, San Joaquin Refining Co., 2 FEA, par. 83,130 (April 25, 1975); or cash flow achieved in a representative year, Pasco, Inc., 2 FEA, par. 83,168 (June 3, 1975). In at least one case FEA granted relief designed to enable a new entrant to achieve a return on invested capital which approximated the industry average, Wickett Refining Co., 2 FEA, par. 83,238 (August 6, 1975).

The relief granted was relief from less than all of the firm's entitlements obligations, usually for a period of three or six months. If data submitted after the expiration of the period of relief established a continuing loss of profits deemed sufficiently harsh to justify relief, it was extended, again usually for three months. Famariss Oil Corp., 2 FEA, par. 83,213 (July 9, 1975); Good Hope Refineries, Inc., 2 FEA, par. 83,214 (July 9, 1975). Eventually, the agency consolidated thirteen such applications for extension of relief. 2 FEA, par. 83,356 (November 7, 1975).

[84] 2 FEA, pars. 83,883-84 (September 11, 1975).

Rule No. 3,[85] promulgated approximately one month after the institution of the program. Second, the agency had to decide whether relief in any form should be continued after the expiration of Special Rule No. 3 and, if so, what degree of hardship must be demonstrated in order to entitle a firm to relief. Finally, the agency had to decide how much relief should be granted if it was decided that small firms should be protected.

The basic policy decisions reflected in these problems had already been made by February 1975; the issue between that time and the formal development of the *Delta Refining* rule was refinement of the measure of hardship and relief. After *Mohawk* but before *Delta Refining*, the agency had clearly decided to grant that relief deemed necessary to help the firm attain its historic operating posture when it was shown that the application of the entitlements program to that firm would result in serious financial hardship. The only issues remaining were the definition of that serious hardship and the computation of the relief to be granted.

Of course, the rulemaking process may also be utilized to determine the type of relief to be granted by the agency in resolving a given problem. For instance, in the crude reseller rulemaking inquiry noted above, FEA requested comments on the appropriate measure of relief for the passthrough of nonproduct costs by crude resellers. Note that even a general rule may take into consideration variations in the circumstances of members of a class:

> FEA further solicits comments as to whether any generalized limitations on the passthrough of increased non-product costs should be adopted in the form of a uniform increment per unit for a crude oil reseller, as had been provided for certain other covered products in section 212.93(b) or whether some other mechanism to permit the passthrough of increases in non-product costs for each crude reseller over those incurred in May, 1973 should be adopted in order to accommodate distinctions in the nature of individual crude oil reseller's operations (including but not limited to, volumes and distances of crude transportation among the various types of resellers).[86]

Although use of the exception process may decrease the amount of "regulation" of the petroleum industry as a whole, utilization of that process, rather than rulemaking, may increase the regulatory burden on individual firms. An argument made in favor of the use

---

[85] 39 Fed. Reg. 43814 (1974).
[86] 41 Fed. Reg. 47081 (1976).

of the exception process to resolve individual problems is that such use reduces the regulatory burden on the industry as a whole by keeping the codified regulations free of numerous exceptions and provisos and by reducing the number of generalized rulemaking proceedings on which firms would feel compelled to comment. Again, there is validity to this point. To the extent that firms in particular situations are aware that the exceptions process offers relief for their specific problems, there is reduced harm in providing for only the generalized rule in the published regulation. The recent publication of more specific guidelines upon which exception relief has been given, as mandated by ECPA, increases the opportunity for members of the industry to learn of the types of individual relief which may be granted.

Nevertheless, the utilization of the exceptions process by each individual firm seeking relief from a specific problem is very burdensome compared with the use of a general regulatory modification meant to take care of the same problem. An exception application requires the presentation of substantial financial information, supported by legal and policy reasoning, to justify the claim for relief. Most importantly, exception relief is generally granted only for a three- to six-month period, and the applicant must resubmit data on a recurring basis to justify the extension of the exception. This information will generally include audited sales and financial data designed to demonstrate that the grounds upon which relief was first granted still exist. A clear example of this situation is shown by the *Producers Pipeline* and *Marvin Boyer* cases.[87] Relief was granted for a period of about six months. If the firms desired to extend their exception and raise their prices to reflect continued cost increases, an additional filing was required:

> In the event that Boyer requests extension of the exception relief granted herein to reflect additional relief on nonproduct costs herein, it shall submit on or before April 25, 1976 the level of nonproduct costs which the firm incurred in each of the above categories for its 1975 fiscal year and the first quarter of its 1976 fiscal year.[88]

This type of detailed financial reporting in order to correct an apparent oversight in the drafting of the FEA regulation governing resellers is burdensome, particularly to small firms. If such oversights are corrected by changes in the regulations, such as FEA is now pro-

---

[87] See above, p. 125.

[88] Marvin E. Boyer Co., Inc., 3 FEA, par. 83,321 (December 1, 1976).

posing to do with respect to crude oil resellers, relief is automatic and continuing, with no filing to FEA required: the firm merely complies with the amended regulation and maintains sufficient records to demonstrate compliance to FEA auditors.

Rulemaking by exception may also have the effect of denying relief to individuals injured only slightly by a poorly drafted regulation. In the case of crude oil resellers, exception relief from the prohibition on the passthrough of nonproduct costs is usually given only when the reseller "is experiencing an economic disincentive to continue its reselling operations as a result of" the restriction.[89] Consequently, there may be resellers who are suffering losses as a result of this regulatory oversight, but who do not request exception relief because they do not believe they can demonstrate that their losses are large enough to cause them to terminate their businesses. Were FEA to have remedied the problem by a rulemaking (as it now proposes to do), all resellers would have been able to obtain relief, not just those about to be forced out of business.

On balance, then, it appears that much of the advantage to be gained by utilizing the exception process to keep the formal regulatory structure simple and straightforward, and to reduce the number of rulemaking proceedings necessary to deal with specific problems, is counterbalanced by the fact that the exception process may operate to place substantial reporting and financial burdens on firms which must rely on the exception process for continuing relief.

Notice of changes in regulatory policy may not be adequate where rulemaking is by exception. FEA and the Office of Exceptions and Appeals have made great efforts to insure that the contents of exception decisions are widely disseminated. The decisions are compiled and published, accompanied by subject headings and cross-reference tables. In spite of this availability of exception precedents, however, and of the fact that the notice of all exception applications is given in the *Federal Register,* the exception process does not generally provide as adequate notice of changes in regulatory policy as the rulemaking process.

It may be argued that the exception process is more efficient because it resolves disputes between specific parties, whereas a rulemaking may affect large classes of firms in the industry. Thus, if relief in cases like *Marvin Boyer* (involving passthrough of crude resellers' nonproduct costs) or *Austral Oil Co.* (involving sale of old oil at upper tier prices to encourage new investments) were proposed

---

[89] FEA, Office of Exceptions and Appeals Guidelines, 41 Fed. Reg. 50856, 50860 (1976).

130

through the rulemaking process, large numbers of individuals would file comments because they would assume, correctly or incorrectly, that their interests might be adversely affected.

In the case of crude resellers, those who purchased through crude resellers might file comments in a rulemaking because they did not wish to see their crude oil costs increase. In the case of production incentives for lower tier oil, producers selling at upper tier prices would voice concern because, given the congressionally set national average price for crude, if significant volumes of lower tier crude are sold at upper tier prices, then the price of upper tier crude must be decreased to compensate for the increased volume of oil now being sold at the higher price. Thus, if the rulemaking process were used, FEA might be faced with major issues of crude oil pricing policy. On the other hand, in the exception process, the specific case to be decided concerns only one crude oil reseller and one lower tier producer seeking additional investment incentives.

The fact remains that a particular application for relief, at first seeming far removed in a practical sense from most of the industry, may establish a precedent for the granting of future relief. *Boyer* and *Austral* may well be relied on by increasing numbers of firms over periods of time. Few firms will have made a presentation in the initial decision of these cases because they had no way of knowing what the "rule" adopted by the case would be until the actual decision was announced, and they might not feel sufficiently involved in the case to file an appeal. But once a rule is established and large numbers of successful applications for the price increases follow suit, the effect on the firms in the petroleum industry is just as great as if the decision had been made by rule.

The task force thus seriously questions the appropriateness of utilizing the exception process to avoid the decision-making complexities or industry-wide uncertainty that may result from having a rulemaking which brings into question a particular existing regulatory provision. That is not to say that it is always inappropriate to make general policy decisions by rulemaking solely due to the notice problem.[90] It is only to say that utilization of the exception process to make major decisions may leave those who are concerned about the rule adopted in the decision without an adequate forum for the consideration of their views.[91]

---

[90] See NLRB v. Bell Aerospace, 416 U.S. 267, 295 (1974).

[91] *Recommendations and Reports of the Administrative Conference of the United States*, vol. 2 (Washington, D.C.: Administrative Conference of the United States, 1973), p. 179.

The exception process, when used to develop rules, may provide too limited a scope of relief as compared with other forms of rule-making. A fundamental problem with the utilization of the exception process as a forum for the development of rules of general application through an incremental process is that the exceptions process does not appear broad enough in conceptual scope to provide the full range of remedies that would be available using other forms of rulemaking.

In essence, the process has developed from a function which was designed to provide exceptions from rules and not to determine the rules themselves. Thus, the exception process inherently focuses on the amount of relief necessary to alleviate a particular "inequity" or "hardship," and has therefore been especially concerned with insuring that the relief granted in a particular case does not create "windfall profits," [92] or place a firm in a profit position more advantageous than it has previously experienced.[93]

Similarly, exception relief is generally granted for short periods of time to insure that relief does not continue after the factual predicate upon which it was granted no longer exists. Thus, the process is not well adapted for the prospective modification of rules that are poorly written or whose logic has been undercut by changed circumstances.

The effect of these limitations of the exceptions process is shown most clearly by the line of exceptions decisions culminating in the *Austral* case. Producer cases such as *Pruet & Hughes Co.*,[94] and *Braden Deem, Inc.*,[95] involved production disincentives that resulted from the tier pricing system. It is important to note that these cases involved increased operating costs resulting in a disincentive to continue the extraction of oil from existing wells at current levels of production.

In these cases, although the crucial factor influencing the agency to grant relief may have been the potential loss to the nation of the continued production of the properties involved, it is also true that the companies themselves were suffering a hardship insofar as they would be unable—without relief—to continue to earn a return on their existing investments. In fashioning the relief to be given in this line of cases, the agency properly recognized the need to avoid the granting of windfall profits by limiting relief from lower tier pricing restrictions to the amount needed to offset demonstrable increases in operating costs associated with extraction of the oil.

---

[92] For example, Austral Oil Co., 4 FEA, par. 83,017 (July 15, 1976).

[93] Oskey Gasoline & Oil Co., 2 FEA, par. 83,360 (April 9, 1975).

[94] 2 FEA, par. 83,270 (August 25, 1975).

[95] 3 FEA, par. 83,072 (January 15, 1976).

However, the need to accept the inherent limitations of the exception process in terms of the scope of possible remedies available is far less clear in cases such as *Austral* itself, involving a firm's alleged lack of incentive to make a *new* investment considered to be in the national interest. Unlike the cases involving a disincentive to continue an ongoing operation, these cases involve firms having the option of considering alternative investment opportunities, since they do not necessarily have a present investment in the proposed project. These firms could afford to await the outcome of a general rulemaking proceeding which would consider the full range of potential issues involved—including the issue of whether any relief at all is warranted.

Seen in this context, cases considering relief to permit the beginning of new investment projects do not appear to be "exceptional relief" cases in the usual sense, in which the problem directly at hand is a hardship or inequity suffered by an individual or a firm. Rather, such cases are attempts to change basic regulatory rules because the rules themselves are tending to frustrate national energy goals. The fact that the change is made piecemeal—that is, on a case-by-case basis—only makes the rule change gradual; it does not change its character to one of special relief.

It is also important to note that where the case is not one involving "exceptional relief," in the sense that a measurable financial hardship is being experienced, the remedy that can be given in an exceptions case will have to be somewhat artificial. Assuming that changing the rule itself is not open for consideration, the remedy must be directly measured by some profit-related test, such as rate of return, if the agency is to insure that windfall profits do not result from the granting of relief. Compare this with cases involving operating cost increases on existing investments, where the amount of relief granted can be related to the amount of increased costs experienced. Moreover, since for new investments the relief will be of a continuing nature, the measurement of actual profitability (rate of return) will have to be made from time to time on a continuing basis if the agency undertakes to insure fully that windfall profits do not result from the granting of exceptional relief. This, of course, changes the very nature of the FEA regulatory program for the companies involved to one of the model of public utility regulation—a model which the task force believes is unsuited to the petroleum-producing industry and certainly not the type of regulation envisioned by Congress under the EPAA, as amended.

In passing EPCA, Congress appears to have made a deliberate tradeoff between controlling inflation on the one hand and increasing economic incentives for crude oil exploration and production on the other.[96] Thus, in setting an average price for controlled domestic crude, with a 10 percent annual escalation provision, it cannot be assumed that the at least occasional appearance of disincentives to produce domestic crude oil would not occur as a fully acceptable result in light of the countervailing objective of inflation control. Note also that Congress, in passing EPCA, recognized that the establishment of several tiers, with a fixed national average, necessarily meant that an increase in the price allowed for one class would require an offsetting decrease in the allowable price for the other classes. This remains true regardless whether the relief is granted through exceptions or changes in the rules themselves.

Nevertheless, it might be argued that the use of the exception process in *Austral* was correct if few producers take advantage of the decision and the volume of old oil produced from new investments at upper tier prices is not large enough to have an impact on the average weighted price of crude oil. Taking this view, the use of the exception process simply permitted an increase in the incentives for production of crude oil from old properties without the necessity of a rulemaking hearing to revise the tier system. It is possible, though, that over the long run, many producers will, indeed, seek relief under the *Austral* precedent. In that case, a rulemaking would be needed to decide formally the allocation of permitted average crude cost increases between developers of old property, new property, and those wishing to introduce enhanced recovery techniques.

On the other hand, if only a few producers take advantage of *Austral*, then it could fairly be said that the existing tier-price system works well: the price of old oil is controlled to prevent the reaping of windfall profits, and only a few, marginal barrels are not produced as a result of the price ceiling. If this result is the actual outcome, the exception process, in deciding the equities of the one application before it, may have begun the transformation of FEA regulation into a system of controls on the public utility model solely to encourage the production of a few thousand barrels of crude oil.

The *Austral* case and its progeny thus clearly demonstrate the dangers of using the exception process as a "cutting edge" for developing general rules when a request for relief is not based on an actual hardship to an applicant but on a desire to further national energy goals.

The use of exception resources to make automatic adjustments is wasteful. Finally, the task force believes that the utilization of the

---

[96] H.R. Rep. No. 94-1392, 94th Congress, 2d session (1976), pp. 55-58.

exception process, as such, for the making of automatic adjustments is wasteful both of the Office of Exceptions and Appeals' analytic personnel and of an applicant's time. Where, as in the case of natural gas liquid regulations, adjustments are to be made automatically to reflect increased costs, without a further showing, there is no necessity for the Office of Exceptions to promulgate an elaborate decision, making findings of fact and conclusions of law according to a routine procedure.

Whenever, through rule or exception decision, FEA determines that relief is to be granted or measured according to a standardized set of criteria, the use of which by a given applicant requires no policy decision on the part of FEA and requires the utilization of few analytic skills, it would seem more sensible for such an adjustment to be made through some process other than the exceptions procedure. FEA manpower would be used more effectively, and the applicants would be spared the need to prepare an elaborate, though boiler-plate, exception application.

The use of the exception process to make rules of general application in a program scheduled to expire has more disadvantages than advantages. From the above analysis, the task force concludes that the problems associated with utilizing case-by-case procedures to make rules of general application outweigh the potential benefits to be gained. In general, case-by-case rulemaking (to the extent that it is otherwise appropriate) assumes a relatively permanent program in which case law may be cumulated over time in order to resolve regulatory questions. FEA regulatory programs, on the other hand, are to expire in 1979 for crude oil, and sooner for products as decontrol progresses. Consequently, the adoption of case-by-case rulemaking as an integral part of the FEA regulatory development process does not appear appropriate at the present time.

More importantly, the type of case-by-case rulemaking presently found in the exception process creates vested interests in the maintenance of the FEA regulatory structure and has established an entering wedge for the adoption of utilitylike rate-of-return regulation, developments which, if left to continue, may set in motion forces which undercut the temporary nature of FEA regulation. These problems, however, appear an inherent part of any attempt to utilize the exception process to make rules of general application, since the exception process does not permit the consideration of the broad range of solutions to a given situation that would be available through formal rulemaking. Nor does the exception process result in as wide participation by the public and industry in the making of agency decisions as does rulemaking.

At the same time, there appear to be few benefits from the use of the exception process in this fashion. Apart from the ability of an exception decision to tailor-make the amount of relief to fit the financial situation of a particular company, there is no inherent advantage in the exception process over a broad rulemaking inquiry as a method of deciding the general type of regulatory response, if any, which is appropriate to deal with a given problem. Likewise, there does not seem to be any overriding advantage to utilizing the exception process to determine the degree to which a particular problem is characteristic of a large segment of the industry; a general rulemaking inquiry can achieve the same result. Lastly, the exception process, with its requirement for frequent renewals of relief and continuing agency oversight, may impose burdens on affected firms that could be avoided by direct amendment of an inappropriate regulation.

In sum, the task force believes that continued use of the exception process as a "cutting edge" for regulatory development, including the establishment of relief policies which serve to maintain the profit levels of classes of firms, is unjustified, and that the essential function of the exception process should be to provide relief for actual hardship or individual inequity.

The Regulatory Development Process*

The task force undertook an in-depth review of FEA's organization and process for development of the petroleum allocation and pricing regulations. The task force determined that the principal problem with the agency's regulatory development process is the absence of a standard procedure. The failure to delineate clearly functional responsibilities has created a blurring of jurisdictions, uncertainty of roles, and limited participation in decision making. Appropriate offices do not always participate in the development process. Issue papers are not always prepared and circulated. Drafts of proposed regulations do not consistently receive attention from offices that should be making contributions.

Three other significant problems exist. First, there are no standard review criteria. Economic analyses are infrequently sought or employed to influence regulatory drafting or decisions. Other criteria, such as statutory goals and enforceability, have not been systemat-

---

* Editor's note: Only the task force's summary of the problem with the agency's regulatory development process has been included.

ically considered by regulation drafters and reviewers. Second, the offices within the agency which are naturally responsible for regulatory development are located in different parts of Washington. This physical separation has handicapped communication and coordination between offices. Third, employee knowledge of the regulations and the industry is inadequate.

4

CONCLUSION: THE COSTS OF THE CURRENT FEA REGULATORY PROGRAM OUTWEIGH ITS BENEFITS

By establishing the authority of the FEA to regulate the allocation and pricing of crude oil and its products, Congress intended to achieve four basic objectives: to keep the high OPEC-set price of crude from determining the price of domestic crude oil and petroleum products when supplies of petroleum are adequate; to prevent the establishment of unreasonably high prices for petroleum products in an embargo or other temporary supply interruption; to allocate scarce petroleum products on an equitable basis during a shortage; and to prevent a temporary supply shortage from acting to decrease the extent of competition existing in the petroleum industry.

In carrying out the responsibilities placed upon it by Congress, FEA has developed a complex, comprehensive system of regulations, which has led to the creation of numerous volumes of rules, interpretations, exception decisions, appeal decisions, guidelines, forms, and explanatory material. In the previous sections of this report, the task force has analyzed the impact of the FEA regulatory structure on the various segments of the petroleum industry, on taxpayers, and on the consuming public.

In this section of the report, we will attempt to balance the cost of FEA regulations as they now exist with the benefits they create for the public. After so doing, we will make a judgment as to the necessity for changing the present system of regulation. Throughout its study, the task force has taken one decision as given: the determination by Congress that the price of domestic crude oil should be controlled until 1979 at a level below the world price of crude oil as set by OPEC. Consequently, the analysis made in this section assumes that these crude oil price controls will remain in effect until that time, and we will weigh the costs and benefits of continuing the present FEA regulatory structure on that basis.

FEA Product Pricing and Allocation Regulations, the Crude Oil Buy/ Sell Program, and the Supplier/Purchaser Freeze Are Unnecessary in Present Supply Conditions. At the present time, world supplies of crude oil and petroleum products, including propane, are adequate. There is also substantial excess refinery capacity, mainly abroad. Under these conditions, there is simply no need for regulations which control the distribution of refined petroleum products and residual fuel and limit their prices.[1] Similarly, with adequate supplies of imported crude oil, there is no reason to maintain the current regulations which freeze supplier/purchaser relationships on crude oil and provide for its distribution through the buy/sell program. In the present supply condition, most products are being distributed in accordance with market needs. As this report has found, the framework for supplier/ purchaser relationships and base period allocations is simply not relevant to today's business relationships. Likewise, prices are being limited by competition, a fact most clearly evidenced by the existence of substantial refiner "cost banks" which, as of January 1, 1976, were in excess of $1.2 billion for the top twenty refiners.[2]

In passing the Energy Policy and Conservation Act of 1975, Congress recognized that in conditions of adequate supply the regulatory structure might be unnecessary. In accordance with procedures set out in that statute, price and allocation regulations have been placed on a standby basis for a majority of products. But regulations remain in force for several other products, principally motor gasoline and propane.

We recognize that there is some concern among branded independent dealers that eliminating the regulations on product allocation would result in their suppliers' terminating their supply relationships. As this report noted, however, when made in a time of surplus, such supplier terminations are generally the result of market conditions and changes in marketing strategy. Congress has made clear that any long-run structural problems which may be present in the petroleum industry are not to be remedied by reliance on the FEA regulatory system but through legislation or antitrust action. We note that legislation dealing with dealer terminations was reported in both houses of Congress during the 94th Congress but was not enacted.[3]

---

[1] See, however, our suggestion with respect to the retention of end-user controls on propane and naphtha.

[2] FEA, *Preliminary Findings and Views Concerning the Exemption of Naphtha Jet Fuel from the Mandatory Allocation and Price Regulation*, August 13, 1976, p. 53.

[3] H.R. 13000, 94th Congress, 1st session (H.R. Rep. No. 94-1615, 1976); S. 323, 94th Congress, 1st session (S. Rep. No. 94-120, 1975).

In summary, this task force finds that present market conditions do not require continuation of present product pricing and allocation regulations, the crude oil buy/sell program (except for northern-tier refiners), and the freeze on supplier/purchaser relationships on crude oil.[4] Only the entitlements program, which complements the statutorily mandated crude price regulations by equalizing refiner's crude costs, is necessary at the present time. As will be discussed further, modification of the entitlements program will result in even greater benefits to the public.

The Current Product Price and Allocation Regulations Will Not Work in a Shortage. Perhaps the most significant deficiency of the current regulatory scheme is not that it is unnecessary in a period of normal supply but that it would not work should another embargo occur. In general, 1972 base periods remain in effect—for implementation in 1976 or 1977. In some cases, however, these base periods have been adjusted. First, certain priority users are entitled to receive an allocation equivalent to 100 percent of their current needs. As these consumers' current needs increase, the increase in allocation levels is relayed back up their suppliers' chain of distribution. Second, the regulations initially provided an automatic "unusual growth" adjustment for increased usage in 1973 over 1972. For a time during 1974, FEA regional offices were also permitted to make adjustments to purchasers based on those purchasers' "changed circumstances." Third, the FEA Office of Exceptions and Appeals has permitted certain purchasers to receive exception relief permitting increased base period usage in order to avoid "serious hardship" or "gross inequity" for those individuals. Lastly, new firms have entered the marketplace and existing firms have switched suppliers. With FEA concurrence, these additions and changes have resulted in the assignment of new suppliers and new base period volumes, at levels potentially in excess of current needs.

The mere recitation of the myriad changes that have occurred demonstrates that the existing system simply will not govern allocation levels for various fuels in the event of a future shortage. Just as importantly, there has been no enforcement supervision of the allocation program since shortages have ended. Rather, allocation certificates on file represent the unverified and unaudited statistics supplied by users to their suppliers and by those suppliers, in turn, to their

---

[4] It should be noted that because crude prices are controlled and refiners are prevented from bidding for domestic crude, the allocation program for Canadian crude may be necessary during the transition by northern-tier refiners from their reliance on Canadian crude to other sources of supply.

own suppliers up the channel of distribution. Moreover, the task force has also discovered that there is no mechanism (other than occasional ad hoc attempts by regional offices) to reduce the allocation of a reseller when the reseller loses customers to another supplier. The reseller who gains that customer in a switch is, however, permitted to obtain an increase, thus resulting in a double counting for the switching customer. It is not surprising, then, that FEA officials believe that the current authorized allocation levels for petroleum supplies are substantially in excess of current usage. Indeed, FEA officials estimate that, for motor gasoline, the allocation entitlements presently on record are 130 percent of current usage.

If this situation were not discouraging enough, the task force has also found that in the current supply condition, many supplier/purchaser arrangements no longer reflect the "official" relationships which are established by the program, but have changed in accordance with the dynamics of the marketplace. Consequently, many base period relationships which legally exist as part of the present program simply do not reflect current market distribution patterns. If, in a time of shortage, the current regulations were to be placed into effect, complete chaos would result as many transportation and distribution patterns would have to change instantaneously and many sellers would be placed at a competitive disadvantage because of artificially low base period volumes. Indeed, the task force has reason to believe that many suppliers and purchasers have not kept records that would be sufficient to identify their base period suppliers and customers in 1972 and what the appropriate base period volumes were. Hence, many purchasers and suppliers would not have the information necessary to establish their "official" relationships even if they have the genuine desire to do so.

Likewise, the price regulations for refiners would be difficult to enforce in a shortage. As of early 1977, FEA had not yet fully audited base period figures for refiners' sales to each class of purchaser in 1973. It is obvious that if base period prices for the last shortage are not yet known, there is still significant uncertainty as to the precise impact of price controls within the current framework during another shortage. Additionally, the current "banking" regulations for refiners are so complex that it is extremely difficult to determine, at any given point in time, the appropriate price ceiling for a particular product. Such ceilings are, of course, of crucial importance in a shortage, but they could not be effectively audited during an emergency period.

The task force has also found that FEA's Compliance Office believes that very few resellers are in compliance with price regulations. Again, if not all retailers have appropriate price records and do not have an audited base period price, it is simply impossible for meaningful price controls to be reimposed on the reseller segment of the industry under the present regulations.

The Regulatory Structure Imposes Substantial Costs on the Business Community. The petroleum industry incurs substantial costs as a result of the FEA regulatory structure, itself. The task force has found that, for example, the petroleum industry must file about 600,000 forms a year with the FEA. Statistics provided by several U.S. refiners indicate that the administrative costs of filing these forms and hiring legal and accounting staffs to insure regulatory compliance may result in costs of as much as $570 million annually for all refiners. Additionally, businessmen must apply to the FEA for permission to engage in normal business transactions. For instance, about 2,000 to 3,000 cases per month were filed with FEA regional offices in late 1975 and early 1976. Of these, approximately 1,000 cases per month concerned assignment of base period suppliers and volumes for motor gasoline retailing, the sector of the industry containing the largest number of small businessmen. Similarly, businessmen may find it necessary to file applications for exception relief. The task force has found that exception and appeal requests have amounted to about 3,000 cases during the last year.[5] Businessmen may even have to file exception applications to change suppliers or to shut down their refineries for repairs.

The FEA pricing structure also creates competitive problems among businessmen. In many cases, businessmen have complained because the regulatory structure ties them to specific suppliers. Although they may wish to change suppliers, their base period suppliers may not wish to lose them as customers; conversely, new suppliers may often be wary of incurring a new supply obligation should a transfer be made formally. The task force finds this lack of flexibility for those who consider themselves independent entrepreneurs to be particularly disturbing. Also disturbing to the task force is the fact that the "class of purchaser" rules have effectively frozen price differentials that may no longer be justified, resulting in artificially low cost levels for independent jobbers and nonbranded distributors. These differentials may have caused forward integration by jobbers into retail service station operations, when such integration might have been economically unjustifiable without the artificial advantage given by the price regulations. Likewise, the branded independents

---

[5] Data furnished to task force by the FEA Office of Exceptions and Appeals.

may often find themselves faced with a severe competitive squeeze because their nonbranded competitors are receiving substantially lower priced gasoline.

The Regulations Impose Costs on Taxpayers and Consumers. The task force has found that administrative costs of FEA which must be borne by the American taxpayers are substantial. During fiscal year 1977, $47 million is budgeted for FEA regulatory activities. Since the inception of the petroleum regulatory program in 1973, almost $500 million has been spent to regulate the petroleum industry. In these times of concern over government expenditures, the maintenance of such a substantial administrative bureaucracy can only be justified under the most compelling circumstances, circumstances which we have shown are absent in the present case. Additionally, the consumer must, of course, ultimately pay the price for the cost of regulations which are imposed on the petroleum industry.

More importantly, the task force has found that the entitlements program, as presently structured, creates an effective barrier against the importation of products. The task force has also found that present regulations discourage the construction of cost-efficient new refining capacity. With effectively cheaper foreign products blocked from importation, and with necessary new capacity unlikely to be available to meet increases in future demand, it is reasonably certain that additional supplies must be obtained by ever-increasing utilization of existing capacity. The likely result will be to raise per unit out-of-pocket costs of production and, thus, to raise consumer prices.

High or rising utilization levels are not a sufficient inducement to increase capacity. Firms are unlikely to invest in the construction of new facilities unless they perceive that the prices they will be able to charge for the output will be sufficient over the life of the facility to cover both out-of-pocket costs and a rate of return on the investment equal to what they could earn by using the money in other ways. The existing regulatory structure prevents full attainment of such a price now and, in addition, does not offer sufficient stability over time on which to base investments. The biggest inducement to increased utilization, the entitlements program, will end when the price controls on domestic crude oil end in 1979. Thus, the incentives work to push up utilization, but not to expand capacity.

By contrast, refineries outside the United States are operating far below capacity. If the entitlements program were structured to grant entitlements for the importation of products, the savings which would result from utilization of relatively idle foreign refineries could

144

be passed on to the consumer. Moreover, retention of fees on the importation of product under the Mandatory Oil Import Program (MOIP), as recommended by the task force, should insure that in the long run new refining capacity which is constructed to meet U.S. demand will be located in the United States rather than abroad.

The small refiner bias also increases the cost of petroleum products. The task force has found that the $400 million subsidy given to refiners under the small refiner bias is far in excess of that given in any earlier program. The subsidy granted by the bias must be paid for by the other refiners. Since it is these refiners who provide the marginal source of supply, the higher prices they must charge to pay for the bias payment set the market price and are passed on to the consumer. With the end of price controls on middle distillates and residual fuels, small refiners are not required to lower their product prices to reflect payments under the bias system. Thus, a large proportion of these payments may now be retained as pure windfall profits.

Insofar as the allocation regulations constrain businessmen from shopping for the most efficient source of supply, consumers are also hurt. The end result is that the American consumer is not achieving all of the savings which should result from the congressional decision to control domestic crude oil prices. Thus, a revised system of entitlements which include products would result in substantial savings to the American consumer.

Continuation of the Present Controls Will Produce Long-Run Inefficiencies. The task force has found that current price controls on refiners when combined with the small refiners bias make investment in small refineries artificially profitable. Thus, the regulations, if continued, will result in the construction of refineries which are below the size which would most efficiently produce refined products. Likewise, the FEA price structure, particularly as it now exists—with middle distillates and residual fuel decontrolled, and gasoline prices remaining controlled—discourages the construction of the more sophisticated equipment needed to make gasoline. Consequently, if continued over time, the petroleum regulations may well saddle the American economy with inefficient refineries and create a shortage of equipment necessary to produce gasoline, particularly nonleaded gasoline.

Over the long run, the task force has found the FEA regulations will also result in the creation of inefficiencies in the distribution of petroleum products. Artificial price differentials unrelated to efficiency may be maintained, thus reducing the incentive to find less

145

costly methods of marketing. Also, FEA regulations may discourage the movement of products among regions creating a potential for regional shortages and regional marketing imbalances.

Conclusions. In summary, we find that FEA regulations as they now exist confer few, if any, benefits on the public. Moreover, though the current regulations are totally unsuitable to manage a future shortage, their existence may lull the American public into a false belief that a program now exists for the management of possible energy crises. In return for this lack of benefits and sense of false security, the American businessman, the taxpayer, and the petroleum consumer must incur higher costs than might otherwise be the case. Indeed, continuation of the present regulatory mechanism will result in long-run inefficiencies for the American economy. It is for this reason that the task force concludes FEA regulations must be changed. In the next section, we will set out our recommendations.

5
RECOMMENDATIONS

Recommended Changes in FEA Regulations

Recommended Changes during a Normal Supply Period. Seven changes in FEA regulations are recommended by the task force for periods of normal supply conditions.

Eliminate all product price and allocation regulations for refiners and resellers. Controls on the sale of product by refiners and resellers are both unnecessary and unworkable in present normal supply conditions. Prices have for some time been constrained by the market rather than by regulations, as evidenced by the growth of cost banks. Future price increases (beyond those to cover increased crude oil costs) can be restrained through crude oil and product entitlements. Additional regulations to achieve these ends only add costs and thereby introduce pressures to raise prices. Allocation regulations are unnecessary except in a shortage; surplus product is generally available so that no refiner or marketer, except one with a very uneconomic location, should be lacking for supply.

Implementation of this recommendation would require an energy action.

Eliminate or phase out the crude oil buy/sell program. Allocation regulations other than entitlements are also unnecessary for crude oil because there is no shortage of crude oil. Both the crude oil buy/sell program and the freeze on supplier/purchaser relations should be eliminated.[1]

---

[1] It should be noted that, because crude prices are controlled and refiners are prevented from bidding for domestic crude, the allocation program for Canadian crude may be necessary during the transition by northern-tier refiners from their reliance on Canadian crude to other sources of supply. No other refiners, however, should be included.

The benefits of the crude oil buy/sell program are concentrated on relatively few firms. According to data supplied by FEA, of 124 refiners who might be covered by the program, 25 receive no allocation and 28 more receive an allocation but do not purchase any of it. Nine firms formerly purchased crude oil through the program, but dropped out when the price regulations governing it were changed. Thirty-eight firms buy only occasionally or only a small amount of their allocation. Only 24 firms remain that buy either large volumes or most of their allocation. Ranked in order of the volume of their purchases through the program, the top three buyers take 51 percent of the total; the top eight buy 81 percent. The largest purchaser was classified as a large independent refiner (refining capacity in excess of 175,000 barrels per day, and producing less than 30 percent of the crude oil it used) in December 1973 when the EPAA was passed; the second largest purchaser has grown to that size since December 1973.

Moreover, the amount of oil involved in the program is very small in relation to the total amount of oil refined domestically. The total purchases represented about 3.3 percent of total crude runs to stills in 1975, and about 4.5 percent of crude runs to stills in PADs II, III, and IV, the regions where most of the buyers are located.

Continuation of the crude oil buy/sell program, except as noted for the northern-tier refineries, serves primarily as a subsidy to those firms that buy through the program. Firms only use the program to the extent that they can purchase crude oil at a price below that which they would pay in the open market. There are three ways in which the buy/sell program may lower the price to the buyers below the market price. First, the price set in the regulations for buy/sell oil (now the average import price) may be lower than the price the buying firm would have to pay for oil in the open market. This could arise particularly if the firm had a need for the more expensive qualities of crude oils. Second, even if the price the firm would pay in the open market were equal to the average import price, the program may offer lower costs connected with purchasing, such as brokers' fees, than the firm would face in the open market. Finally, the buy/sell program may offer savings to the buying firm on transportation costs if the crude oil actually transferred is domestic crude oil rather than imported crude.

The first two kinds of subsidies, those arising because either the actual price or the purchasing expenses are lower through the buy/sell program than in the open market, certainly existed before the buy/sell price regulations were changed. In March 1976, the price charged for buy/sell crude oil was changed from the average crude oil price (domestic and foreign) plus a thirty-cent handling fee to the average import price plus the handling fee. As a result, nine firms that had accounted for 17 to 18 percent of the total purchases dropped out of the program.

The third way the buy/sell subsidy may work is by lowering the cost of transporting crude oil to the buyers compared with the cost if crude oil were bought on the open market. This can occur only if the crude oil actually delivered comes from domestic wells rather than from imports, as buying firms are required to pay the cost of transportation actually incurred. If it is impossible to transport imported crude oil to a buying refinery, the buy/sell program subsidy could be seen as very large indeed, for it would be serving as at least a temporary substitute for building pipelines to connect with importing centers.

In fact, actual sales under this program between June–August 1975 and June–August 1976 have been composed of between 39 percent and 49 percent imported oil, with an overall average for the period of 43 percent imported oil. The refineries receiving this oil are clearly located adjacent to adequate transportation facilities for importing oil directly. Moreover, the task force has been unable to uncover any evidence that any refiners, other than some northern-tier ones not covered by this proposal, would be unable to acquire imported crude oil due to transportation difficulties. If FEA has further evidence, however, that the transportation facilities for moving imported crude oil are lacking for some refiners, the crude oil buy/sell program could be phased out by 1977 or 1978. FEA should give advance notice if it intends to phase out the program in order that such refiners as may lack access to imported crude will make the necessary arrangements to receive crude oil through alternative means.[2] Implementation of this regulation would require an energy action.

Eliminate the freeze on supplier/purchaser relations for crude oil. The freeze on supplier/purchaser relations for crude oil is unnecessary to guarantee availability of oil during a period of normal supplies. Continuation of the freeze works to block any changes in the distribution of domestic oil from the pattern existing in 1972, despite major

---

[2] It should be noted that the problem of insufficient transportation facilities for moving imported crude oil to some inland refiners is not a problem created by either the Arab oil embargo or by the subsequent regulation of the oil industry. In effect, the regulations, including both the buy/sell program and the freeze on supplier/purchaser relations for crude oil has served as a reprieve for these refiners. It is questionable whether such reprieve is consistent with congressional intent that the regulation not be for naturally occurring shifts in the petroleum industry.

changes in the total amount of domestically produced crude oil. The freeze prevents the market from redirecting reduced supplies to the most efficient refiners, and hinders market changes in the disposition of crude oil supplies in response to changing demands for different types of crude oil. Preventing any change in the disposition of crude oil supplies to reflect changes in both the quantity of available oil and the equipment mixes of the receiving refineries prevents consumers from getting refined product at the lowest possible cost.

The task force has considered whether elimination of the supplier/purchaser freeze for crude oil would result in payments for domestic crude oil supplies that would be in violation of the price rules. The freeze on supplier/purchaser relations is of limited effectiveness, however, as a means of enforcing compliance with the crude oil price control regulations. Abolition of the freeze would permit producers to change the destinations of their crude oil supplies. However, in making the shift producers are limited by the entitlements program in their ability to induce refiners to pay illegal prices for the oil. If there were no entitlements system, then indeed the inducements to refiners to comply with requests for illegally high prices would be large, as lower tier oil is roughly seven to eight dollars less than imports on average. In exchange for access to lower priced domestic oil, refiners absent entitlements would be willing to split that cost savings with producers. Entitlements, however, change the magnitude of the potential differential significantly, although the program does not completely equalize crude oil costs for refiners.

Without the supplier/purchaser freeze, some current purchasers of domestic oil would have to turn either to other domestic suppliers or, more likely, to imports. The only reason that a refiner in such a situation would be willing to pay illegal prices for domestic crude oil would be if either of these alternative sources of supply (a different domestic source or imports) cost him more than the average cost for either of these sources. If the imports available to him were more expensive than average, he would not be fully compensated for his additional expenses by the entitlements program. Thus, he might be willing to make under-the-table payments up to the amount by which his import costs differed from average import costs. If the payments demanded were any larger than this, however, he would be better off buying the more-than-average-priced imported oil. Similarly, if the refiner with entitlements could not get an alternative domestic supply at a price as low as previously, and if transportation barriers prevented him from doing better still by importing, then he might be willing to make such payments. But the amount would be limited to the difference between the price of his former supplies and his next best alternative.

Given entitlements, the difference between a specific refiner's cost and the average cost of acquiring crude oil represents the maximum that refiner would be willing to pay above the legal crude oil price. Study of the banked costs of sixteen large refiners indicates that average crude oil costs net of entitlements in January 1976 varied by $3.26 per barrel. When the highest and the lowest average crude oil costs were removed, the range was significantly smaller; namely, $0.93 per barrel.[3] These figures represent the difference between the top and the bottom average costs; the average is in the middle. Thus, entitlements reduce the incentive to comply with illegal prices to considerably less than the difference between domestic and imported prices. Maintaining the supplier/purchaser freeze on all domestic crude oil in order to prevent a small amount of fraud seems unnecessarily costly.

Elimination of the supplier/purchaser freeze would require an energy action.

Change the crude oil price regulations and the entitlements program to eliminate fixed quality differentials. Crude oil pricing regulations have set the prices of lower tier crudes at their May 15, 1973, levels plus a fixed increment and upper tier crudes at their November 1, 1975, price minus a fixed increment. World price differentials among the various grades of crude have widened since then, but the differentials among the various grades of price-controlled domestic crude have remained fixed.

This freezing of prices in conjunction with an entitlement, the price of which is based on the difference between the average of the prices of various grades in each tier, places some grades of domestic crude, particularly high-sulfur crudes, at a competitive disadvantage. The disadvantage occurs because the price of the corresponding grades of imported crude have risen less than the average price of imported crude. Thus, the price of lower grade imported crude is now below the effective price of the equivalent domestic grade, which is determined by the fixed differential from the controlled domestic price plus entitlements cost.

This effect, while not resulting in massive underutilization of domestic crude oil, does go against the policy objective of reducing

[3] ICF, Inc., *Final Report: Banked Costs and Market Structure and Behavior*, June 1976, p. 11.

reliance on imports.[4] It also can be corrected while meeting the mandates of the EPCA by changing the price and entitlements regulations. FEA would continue to require certification of domestic crude oil as lower or upper tier, as at present. The exact price of such oil would not be set by regulation, however. The entitlements program would instead be used to set the price, while still ensuring that the composite domestic price fell within the limit established by the EPCA. To do this, the entitlements program would require some changes. Instead of calculating the entitlement price after the data on actual sales are in, as at present, FEA would publish, by the last day of a month, the price of an entitlement for the next month. The value would be determined by estimating both the cost of imported crude for the next month and the expected ratio of lower tier to upper tier crude oil.

Once the price of an entitlement were published, the relative prices paid for domestic crude oils of different qualities would be determined by the market. Buyers would bid for the various crude oils based on their calculations of the cost of refining, the cost of transportation, and the like. The need to acquire an entitlement for each barrel of old oil and a fraction of an entitlement for each barrel of upper tier oil would keep the average lower tier and upper tier prices roughly at the levels necessary to achieve the statutory composite domestic crude oil price. Refiners would not bid more for domestic crude oil of a given quality than an amount which, when transportation costs and the entitlement price were added, would equal the delivered price of imported oil of the same quality. Thus, for example, if crude oil of a given quality could be delivered to a refiner for $13.50 per barrel and if the entitlement price were $8.00, that refiner would be unwilling to pay a domestic producer of lower tier crude oil more than $5.50 minus transportation charges for the same quality of crude oil. Raising the entitlement price to $8.25 would lower the price that same refiner would be willing to pay a domestic producer by $0.25.[5]

---

[4] It also lies behind the inability of those producers of high-sulfur Texas crude oil, bound by royalty contracts to sell at the ceiling price, to sell all they produce, which, in turn, may have influenced the Texas Railroad Commission to decrease the allowable production for December 1976.

[5] Adoption of this recommendation would also ease somewhat the dilemma of how to price Alaskan crude oil. FEA must determine the tier to which Alaskan crude oil production will belong. Moreover, under present regulations FEA would have to set a price within this tier for Alaskan oil because there is no comparable field. Adoption of this recommendation would permit FEA merely to decide the tier, but leave it to the market to set the differential between the average price of that tier and the price to be paid for Alaskan crude oil.

Essentially, a free market allowing normal business practices would prevail within the targets established for each tier. Windfall profits would be prevented by the need to buy entitlements. However, no oil would be left unsold because the government would no longer set relative crude oil prices that were artificially out of line with relative world prices for different qualities.

The first month, FEA could use the entitlement price calculated from actual data in 1976 as the estimate, as this price remained fairly constant over the last six months of that year. If there were fore-casting errors, FEA would adjust the next month's entitlement price, thus changing the prices refiners were willing to pay for domestic crude oil. Since it is necessary to make similar adjustments under the present system, this proposed change adds no uncertainty about achieving the statutory composite price.

Restoring a market system for setting the relative price differen-tials among different qualities of crude oil would also reduce still further any need for the crude oil buy/sell program. Refiners who currently own higher grades of crude oil profit under the present regulatory structure by keeping the oil for their own refineries because such oils are relatively underpriced. Having the differentials set by the market would eliminate this incentive to keep the oil.

An additional benefit to be gained by letting the market establish the relation between prices of different qualities of crude oil is to eliminate the potential problem of refiners paying prices for domestic crude oil that might be illegal under current price control regulations. Adoption of this recommendation would mean that changes in the relative prices of various quality crude oils would be legal, unlike the present situation.

FEA may implement the recommended changes to the price regu-lations and the entitlements program by discretionary amendment.

Broaden the entitlements program to include imported products as well as crude oil. Crude oil price controls are expected to remain in effect until 1979, when their imposition is no longer mandatory. Crude oil price controls were enacted with the intent of holding the prices that consumers must pay for refined oil products below world levels. An entitlements program is necessary in times of crude oil price controls in order to prevent some refiners from gaining an unfair cost advantage vis-à-vis other refiners.

The present entitlements program does not give consumers the full benefit of the crude oil price controls. A mechanism is needed that would pass the full amount of profits withheld from producers through to consumers in the form of lower prices. The simplest change in the regulations that would accomplish this goal would be to extend the entitlements program to cover all imported products and crude oil, instead of just crude oil and some residual fuel oil, as is done presently. This extension would ensure that consumers would benefit from international competition in times of worldwide excess refining capacity, instead of insulating the domestic industry from such competition.

Because there is not enough crude oil from domestic wells to fill domestic demand for refined products at the present prices, imports of either crude oil or products are required. If there were entitlements for products, as well as for crude oil, only the different economic costs involved in the importation of crude oil versus refined products would govern the choice between the forms that imports of oil would take.

Given workable competition in the domestic oil industry and a properly functioning system of crude oil and product entitlements, the division of imports between crude oil to be refined in the United States and products refined abroad would be achieved at lowest cost. Whenever refined product from abroad sells for more than it would cost a domestic refiner to import crude oil and refine the product, the proportion of crude oil imports would rise, and product prices would fall. Conversely, if it cost less to import refined products than to import the crude oil and refine it domestically, imports of crude oil would fall and imports of products would rise.

The product entitlements program would eliminate the advantage which domestic refineries have over foreign refineries due to U.S. crude oil price controls. In this respect, the proposal is neutral as to the location of new refineries. Domestic refineries, however, are more expensive to run than foreign ones, in part because of higher costs for such things as land and labor, and in part because domestic refineries have been equipped to produce a larger proportion of the higher valued products from a barrel of crude oil. To the extent that it is national policy to encourage the construction of domestic versus foreign refineries, tariffs and fees on refined products, rather than entitlements, are desirable to achieve this goal. Tariffs and fees are desirable for this policy goal because their level is set by the U.S. government, in contrast to the present level of entitlements protection which depends on the gap between world and domestic crude oil prices and is thus set in part by OPEC.

It is possible for some product prices to rise if entitlements are extended to imported product, but the circumstances under which this would occur are quite limited. If the price of imported product, following the granting of product entitlements, is less than the price of domestic product in the absence of product entitlements, then price will clearly fall. If the reverse is true, then there will be no imports of that particular product, and it is possible for its price to rise if two further conditions exist. First, total imports must increase sufficiently to reduce the number of entitlements allocated to domestic refiners. Some entitlements are already allocated to importers of residual fuel oil, which constitutes roughly two-thirds of all refined-product imports currently. Imports of refined products other than residual fuel oil have been roughly 8 percent of all imports, crude oil included. The result of adding this amount to the import pool eligible for entitlements is not likely to increase significantly the incremental costs that refiners would incur because of receiving fewer entitlements.[6] Second, for product prices to rise because of product entitlements, the incremental cost increase due to the greater proportion of imports must be greater than the decrease in incremental cost that will occur due to lower utilization of domestic refineries. Only if both these conditions are met will there be a rise in the price of the particular products that are not constrained by imports.

Product entitlements also confer two additional benefits. One is a potential price reduction in inland regions; the other is a potential expansion of inventories. Prices in inland regions will be at least as low as they were prior to the extension of entitlements to cover product imports. Prices may fall in some inland regions, moreover, if the coastal price net of entitlements, plus the cost of transportation, is less than the prior inland price.

Product entitlements may increase incentives to expand inventories of certain products above previous levels during seasons of off-peak demand and this increase would cause peak demand prices to be lower. Such a condition would occur if foreign prices fell during the off-season, and if the resulting cost of importing product with entitlement, plus the storage cost, were less than the expected seasonal price. If storage capacity were to be expanded, the resulting seasonal price would presumably be lowered.

The proposed program of price controls on domestically produced crude oil, plus entitlements that apply to both crude oil imports and product imports, if administered without biases toward particular groups, would result in the lowest possible prices to consumers.

---

[6] Increasing the amount of imports eligible for entitlements lowers the ratio of domestic to foreign oil, and thereby reduces the number of entitlements each refiner receives. This is true whether the increase in the proportion of imports results from an expansion in the coverage of product imports or from a decline in domestic crude oil production that is offset by increased imports of crude oil.

Expansion of the entitlements program to refined products could be accomplished by amendment to the regulations.

Eliminate or reduce the small refiner bias. Small refiner biases in the entitlements program have the effect of raising the cost of crude oil to large refiners whose costs are key determinants of the final price of refined products. They also encourage a long-term trend toward the use of less efficient small refineries, thus increasing the average cost of refining crude oil. Both of these effects are ultimately borne by consumers. The subsidy which is created by the bias to the small refiner exceeds anything that is arguably necessary to maintain the competitive viability of small refiners as a class. For these reasons, such biases should be eliminated or reduced drastically. Moreover, in a decontrol situation where small refiners can retain the amount of the subsidy which is greater than necessary to reduce their prices to the market price, the program generates substantial windfall profits.

It is impossible to conclude that no refiner will go out of business if the small refiner bias of the entitlements program is eliminated, or even reduced to the level at which it existed during the period when the Mandatory Oil Import Program was effective. Some small refiners may go out of business, especially those whose refineries were built solely for the purpose of collecting entitlement money. Most of these, however, are small, privately held companies which do not report to the Securities and Exchange Commission, so that it is difficult to establish what fraction of their profits results from the sale of entitlements.

An alternative means of estimating the effect of the small refiner bias on the survival of small refiners is to assume that those refiners that were in business at the beginning of 1974 were in business despite the lack of a small refiner bias, and would continue to be in business today, barring a change in economic conditions, in the absence of any bias greater than that which existed under MOIP. This assumption is not strictly true, as some of these small refiners might have been near to going out of business in 1974 and might have been saved by the regulations. Nonetheless, this assumption gives a rough estimate of the upper bound of the number of refiners that might go out of business with a diminution of the small refiner bias.

Therefore, the task force has tabulated the small refiners which began refining operations after January 1, 1974.[7] There are thirteen

---

[7] FEA, Office of Regulatory Programs, Office of Crude Oil Operations, *Crude Oil Buy/Sell Refineries Capacity List*, September 1975, September 1976.

such refiners in the continental United States, with total capacity of 145,630 barrels per day, or 0.9 percent of domestic refinery capacity.[8] Although these refiners may have come on stream for a valid economic purpose, they are the refiners that are most likely to be threatened by a reduction or elimination of the bias. Since the number of refineries in this situation and their share of total capacity are small, the impact of the diminution of the small refiner bias would appear to be marginal.[9]

FEA may eliminate the small refiner bias by amendment to the regulations.

Eliminate tariffs and fees on crude oil: revise the MOIP to eliminate its exemptions. Tariffs and fees are currently collected on imports amounting to twenty-one cents per barrel for crude oil and sixty-three cents per barrel for products. The crude oil tariffs and fees were initially adopted to protect domestic production from low-cost foreign competition. In the present circumstances of high foreign prices, they are no longer necessary for this purpose. Nor, at their current levels, do they serve to discourage the consumption of foreign crude oil.[10] Consequently, the only effect of import fees for crude oil is to raise prices to consumers. These tariffs and fees should, therefore, be eliminated. If it is felt that a license system is necessary as a means of requiring the reporting of imports for statistical purposes, then license fees and/or tariffs should be maintained at a nominal one cent per barrel.

Product fees and tariffs are designed to encourage the construction of domestic refineries by eliminating the cost advantages enjoyed by overseas locations. If our recommendation for the institution of product entitlements is adopted, the barrier to foreign competition provided by the entitlements program will disappear. Consequently, some system of import fees or tariffs on product imports will be necessary to encourage the continued construction of domestic refinery capacity.

---

[8] Guam Oil and Refining Company, with a refinery in Guam, was not counted.

[9] The analysis assumes no changes in economic conditions that may have made the survival of small refiners less likely in 1976 than in 1974. If such changes have occurred, elimination of the bias may result in more closures than suggested here. If a change in economic conditions has worsened the efficiency of many small refiners relative to larger ones, it is not clear that Congress intended the consumers to subsidize this inefficiency indefinitely.

[10] A higher level of crude oil fees, however, can serve the function of reducing U.S. reliance on imports of crude oil.

At the same time, the existence of fees would not negate the benefits to consumers of product entitlements. A fee system can and should be administered by FEA.[11]

The current MOIP fee system is needlessly complex. Because of changed conditions in the oil industry, particularly the premium for imported oil over domestic, it is now possible to eliminate the exemptions in MOIP. We recommend, therefore, that MOIP be simplified so as to require an across-the-board fee on imported products. The appropriate level of the fee to be charged is beyond the scope of this report. We note favorably that, as part of its MOIP review proceeding, FEA has undertaken to determine this level.[12]

Modification of MOIP would require a presidential proclamation. Implementation of a products entitlement program could be achieved by amendment of the regulations.

Recommended Changes in the Regulations for a Shortage. As has been shown, maintaining regulation of the oil industry during a period of normal supply conditions creates significant costs for the economy in terms of lost production or more expensive production that would have occurred without controls. Regulations are appropriate only when the market cannot adequately cope with whatever basic change has occurred in the forces of supply and demand. In general, changes that result from market events or irreversible geologic factors are best left to the market to manage. On the other hand, market responses to changes in supply conditions which are seen as purely temporary but which may last more than several months are likely to impose large transfers of income from consumers to producers without generating offsetting increases in supplies to justify the transfer. Moreover, such conditions may cause permanent and undesirable changes in the competitive environment. In such situations, regulations may be a more equitable response. Nevertheless, the task force recommends certain changes in the regulations.

Adopt a trigger to commence and terminate controls. Given that not every change in the forces that underlie the oil market should provoke government regulation, a trigger is needed to determine when to impose controls. The definition of "severe energy supply interruption" as used in EPCA is a good one and ought to be used to define the events that set controls in motion. This definition states

---

[11] The task force takes note that in late 1976 FEA had under way a proceeding to determine whether product fees should be suspended for the duration of the entitlements program. 41 Fed. Reg. 30058 (1976).

[12] Ibid.

that the term "severe energy supply interruption" means a national energy supply shortage which the President determines:

> is, or is likely to be, of significant scope and duration, and of an emergency nature; may cause major adverse impact on national safety or the national economy; and results, or is likely to result, from an interruption in the supply of imported petroleum products or from sabotage or an act of God.[13]

One of the consequences of any given triggering mechanism is that the mechanism itself may cause a modification of industry behavior. An example of a triggering mechanism on a standby regulation that may induce changes in behavior is found in the recent agreement between FEA and Congress on middle distillate decontrol, which provides that a price change of more than a minimum amount would lead to some action by FEA. To avoid the burdens of controls, refiners may well decide to keep their prices below the free-market level, in other words, to behave differently than they would in the absence of that particular trigger.[14] The definition of a severe energy supply interruption ensures that a price increase per se will not induce the onset of an officially declared shortage.

It is also necessary that an automatic termination be provided for, so that controls do not continue to create problems after the supply emergency is ended. While FEA has the discretion to provide for the expiration of controls, we propose that FEA consider asking Congress to enact a provision requiring that controls end automatically three months after their imposition unless the President again finds the existence of a "severe energy supply interruption."

The simple trigger defined here minimizes the need for monitoring the industry. The only information that must be monitored continually is that concerning the volume of imports and overall supply and demand. An embargo which is sufficiently severe to induce the President to declare an emergency would be sufficiently important to become known instantly, monitoring or no monitoring.

FEA could adopt this recommendation by discretionary amendment to the regulations.

Amend the price and allocation regulations. Once it is determined under the proposed trigger that imposition of controls is necessary, a number of possibilities for dealing with a severe supply interruption

---

[13] EPCA, section 3(8).

[14] Interview with David Wilson, FEA principal deputy general counsel, October 7, 1976.

are available. While the shortcomings of price and allocation controls in a surplus situation have been described above, EPCA assumes that such regulations are the first choice for managing future shortages. It is likely, therefore, that such regulations will form the basic framework for any future shortage management scheme.

The proposed price and allocation regulations that should exist in standby status for use in the event of a severe supply interruption are designed to work *only* in a shortage situation of that nature. Once the supply interruption was over, continuation of the program would create intense pressures for its abolition, as normal supply conditions would make inequitable many of its provisions. An important quality of the proposed changes is that these same provisions would not be inequitable in times of severe supply interruptions.

The proposed changes would work best if adopted as a whole. Partial implementation may, in fact, be counterproductive. In their entirety, the proposed changes would result in a simpler regulatory scheme, with consequent savings in the costs of regulation paid by consumers and the industry, and significantly greater ease of enforcement by FEA. To achieve maximum effectiveness, it is imperative that when the program is adopted there be an advance rulemaking procedure to educate industry and the public about the regulations and their benefits.

1. Update base period for price regulation: The base period for calculating product prices for refiners and resellers should be recent with respect to the shortage. Continued reliance on a base period remote in time fails to take into account changes in both overall costs of doing business and structural changes in the industry that have occurred subsequent to the earlier base period date. The base period chosen, however, should precede the period just prior to the onset of the severe supply interruption in order to avoid any anticipatory price reactions that involve speculation on the possibility of a shortage. Thus, for example, FEA could use as a base period the fourth month prior to the finding of the shortage and the authorization of the imposition of controls.

This recommendation could be implemented by amendment to the regulations.

2. Require a single maximum lawful selling price within a PAD district for each product of each refiner, permit full passthrough of both product and nonproduct costs for refiners, and eliminate cost banks: A refiner's maximum lawful selling price should be based upon the weighted average price of each product it sold within each PAD district during the base period, including sales at reseller out-

lets owned and operated by the refiner. Full passthrough of all product and nonproduct cost increases since the base period should be allowed, distributed proportionally among products according to the refinery mix. However, increased crude oil cost passthroughs should be permitted only at the time the products are refined, rather than at the time the crude oil is purchased. Stored products should be sold at prices which reflect both the cost of storage and an average inventory cost based on the original cost of production.

In a true shortage, restrictions of price increases to reflect cost increases will be effective as a means of controlling refiner windfall gains, as the market price for products will far exceed the allowed price. Refiners should not be allowed to justify price increases for present output on the basis of unrecovered increases in costs of producing output sold in prior accounting periods. If the prior accounting period was not during the shortage, such a banking provision would mean refiners would be using the shortage to recoup losses brought about by normal market conditions. If the prior accounting period were during the shortage, there would be no unrecovered cost increases.

Inventory costs should not be confused with banked cost. In a free market situation, refiners build inventories of products during the off-season, such as is usual for distillate fuel oil, gambling that the price during the period of peak use will repay the cost of storing and holding the products. When a shortage is expected to last long enough, refiners must have adequate incentives to build inventories of out-of-season products. Full cost passthrough rules would allow recovery of both the cost of storage and an average inventory cost based on the original cost of production so the necessary incentives would be maintained. But note that these are all costs incurred in the production of the products being sold, rather than unrecovered costs of producing output that has already been sold.

A major advantage of such a price rule is its ease of enforcement relative to the rules in effect today. Calculation of the lawful price can be made on the basis of records already being kept for normal accounting purposes.[15] Moreover, only one audit of a refiner's books would be necessary to verify the base period weighted average price.

---

[15] We understand that one integrated refiner has proposed adopting a new pricing system whereby resellers are charged a price for gasoline, plus a charge for transportation, plus a charge for marketing services where these are provided. If such a system were to be adopted, the gasoline prices charged in these sales would be used in computing the uniform refiner price, while the transportation and marketing charges would be frozen at the levels existing during the base period.

Justifying increases in price by showing increased costs according to output mix would also involve resort to records maintained in the normal course of business.

FEA could eliminate banks by amendment to the regulations. Allowance of full passthrough could generally be by amendment but, to cover all eventualities, implementation of the recommendation might require statutory changes.

3. Freeze resellers' margins, allow full passthrough of increased product costs, and eliminate cost banks: Each reseller's margin should be frozen at its average level during the base period. Full passthrough of any increased product costs should be permitted automatically, but not increases in nonproduct costs. If there is a need to adjust resellers' margins to accommodate either changes in total volume sold or changes in nonproduct costs, this should be done on a uniform basis through a rulemaking by FEA, rather than on a case-by-case basis. Reseller cost banks should not be allowed. Enforcement of this price rule would be straightforward. After verification of the base period average margin, the maximum lawful selling price could be determined from an examination of the reseller's invoices for products.[16]

This rule would not apply to refiner-owned-and-operated outlets, which are covered under the pricing rules governing refineries. Those outlets would be required to charge the refiner's regional price, as the average system-wide cost of operating those outlets (the costs covered by a margin) would be included in the uniform regional average price computation of each refinery.

Consideration has been given to the likely effect of this recommendation on equity between different kinds of resellers, given that they will be charging different prices for refined products. The problem of equity between reseller operations that are directly run by refiners versus those operated by independents would only arise if there were a normal supply of product: that is, if at the existing price for refined products there were more than enough supplies to

---

[16] The reseller price rule proposed here has the advantage of being easier to enforce than the existing set of regulations. While a uniform national retail price would be even simpler, such a measure would raise many serious problems. Retail outlets require greatly different numbers of transactions to receive supplies from the refiner: some receive it directly after one sale, others indirectly after several intermediate sales. Outlets in different locations face different costs for transportation, overhead, and labor. A uniform national price, if set high enough to cover the most expensive locations and longest distribution chains, perhaps would allow some retailers to earn substantial windfall profits. If the uniform national price were set to cover average costs, some localities would lose some independent retailers as the allowed margin implied in the price would not be high enough for them to stay in business.

fill demand. As long as there is a severe supply interruption, and as long as the response by the government is to control prices, the fact that different resellers charge significantly different prices for their products will not result in any loss of sales. All the available product will be sold at the allowed prices; the only difference would be in the timing of the sales. Those who had to charge higher prices might find their allotted supplies being sold later in the month than those with lower prices. However, both groups will find themselves with less supply than they could sell at the allowed price. Monthly revenue thus will not suffer because of this price rule; it will suffer because of the shortfall of supplies. In addition, as discussed in the next subsection, the disparity in prices that would occur under this price rule may be much smaller than the discrepancies found during the last embargo when the price of crude oil available to different refiners was not even roughly equalized.

Consideration has also been given to whether the incentives this rule might create would encourage refiners to close down many company-owned-and-operated reseller operations. Some outlets have very high costs, and requiring a uniform refiner regional price will leave some refiner-owned outlets with much higher costs than revenues. The operation of the proposed refiner price controls, however, would include the costs associated with operating all reseller outlets, since the prices charged at those outlets would be included in the weighted average regional price being charged by refiners during the base period. Thus, transportation and other costs of running reseller operations, including those at distant locations, by integrated refiners would already be included in the single maximum lawful regional selling price of refiners. Therefore, while it is correct that some refiner outlets might not strictly be covering costs during the period of controls, those costs would be covered by the price charged to other customers.

The alternative possibility of granting refiner-owned outlets a margin will generally not fully equate costs and revenues of all such outlets because the margin would only cover the average outlet's costs. In fact, it is likely that any incentives to close refiner-owned stations would be the same if a uniform refiner regional price were used as they would be if an additional margin were used. Furthermore, considering the goodwill that would be lost, it is also likely that any such closings would be limited to operations that were economically marginal even during normal supply periods.

In sum, the proposed refiner price rule will ensure that all costs of operating retail outlets are covered on a system-wide basis. Addi-

163

tionally, the fact that these are temporary regulations, designed to end with the end of the supply interruption, ensures that the incentive effects will not outlive the period of interrupted supply. Thus, the proposed changes are fair, and they are likely to be perceived as fair by the industry.

Implementation of these recommendations could be accomplished by amendment of the regulations.

4. Eliminate class-of-purchaser price rules: Class-of-purchaser rules may also be eliminated without harm. In a true shortage, refiners will be able to sell all of their production, at whatever price is allowed by the cost passthrough rules. If a refiner should charge different prices to different purchasers in the same class, or should vary the differentials among classes, it would make no difference to the purchasers. A purchaser will merely pass through the full increase in product price to consumers because in a true shortage the reseller will be able to sell all the product he gets without regard to historic price differentials. Independent marketers, moreover, will be able to maintain their traditional retail margins even if suppliers eliminate discounts, since in a shortage situation marketers would be able to sell their entire allocation even at increased prices.

The only effect of this change would be to cause some change in historic price differentials among market areas, depending upon whether the major marketers of a region have been branded or non-branded stations; consumers who patronize independents may experience relatively higher prices. Those regions, however, where nonbranded stations have had the market edge have been, in part, the beneficiaries of gasoline supplies that were surpluses. These surpluses were sold at discounts that reflected the need to get rid of them rather than any true cost relationship. During a shortage, there would be no such surpluses and, therefore, no market justification for the abnormally low prices.

This effect would be mitigated, however, by the fact that a high proportion of the supplies of independent marketers come from independent refiners, whose costs would not change relative to those of integrated firms. Furthermore, price differentials in a shortage under the regulations proposed here will be nowhere near as great as those that existed during the Arab embargo of 1973–1974. In that situation, there was no entitlements program to narrow the variation in crude costs among refiners and some independent marketers wound up selling gasoline at prices fifteen to twenty cents higher than branded stations. With entitlements, uniform refiner prices within a PAD, and a uniform reseller margin, retail prices will vary only a little.

164

The only abnormal feature is that prices of independent marketers who buy from major refiners will not exhibit their traditional discount below branded prices. In any event, traditional price differentials would be expected to return after the shortage situation ended.

Analysis of available data suggests that there is only one region of the country which is more dependent on independent marketers than other regions. The average market share in all fifty states of sales by independent, non-refiner-affiliated marketers is 9 percent. In the following eight states, such marketers held between 11 and 17 percent of gasoline sales in 1975: Arizona, Idaho, Nevada, New Mexico, Utah, Wyoming, Montana, and South Dakota. Given the small size of this region, it should be seriously questioned whether the harm to consumers in these states due to eliminating the class-of-purchaser rules is worth the nationwide cost of the rules.

With the elimination of class-of-purchaser rules, refiners will be able to vary their prices among regions to solve supply/demand imbalances much more effectively than if forced to follow class-of-purchaser rules. As will be noted below, supplier/purchaser rules prevent the squeezing of independent marketers during a shortage. In addition to being redundant for this purpose, class-of-purchaser rules may, if present experience is any guide, be impossible to administer or to enforce in conjunction with price regulation and should therefore be eliminated.

FEA may eliminate the class-of-purchaser rules by amendment to the regulations.

5. Allocation downstream from the refiner level should be based on an updated base period: The allocation of products should be based on an updated base period, rather than one that is fixed some time in the past. For example, FEA can use as the base year the first twelve months of the fifteen months preceding the authorization of the imposition of controls, choosing appropriate months or quarters as base periods. Clearly, the more current the base period, the less severe will be the displacements caused by the imposition of allocation regulations including a freeze in supplier/purchaser relations.

This recommendation could be implemented by amendment to the regulations.

6. Retain the crude oil buy/sell program without the small refiner exemption: A crude oil buy/sell program would be needed during a shortage. When crude oil shortages are seen as temporary, refiners who were suddenly cut off from their usual sources of supply would have a very hard time bidding away crude oil from other refiners, even if there were no price controls. Price controls would

make it impossible. Domestic refiners who rely on imported crude oil would then be forced to run with substantial excess capacity, creating arbitrary competitive imbalance among refiners, depending on relative access to domestic crude oil. In the extreme, refiners would be forced to shut down. As the shortage is assumed to be a temporary one, forcing refiners to shut down may impose losses on society at large, as well as on the refiners so affected. A continued buy/sell program would thus be necessary to equalize capacity utilization among firms.

The buy/sell program should be redesigned, however, to eliminate the present structure under which all but the fifteen largest refiners are exempted from obligations to sell and small and independent refiners are the sole beneficiaries of the rights to buy. In a shortage, small refineries may therefore be able, on the average, to run at higher levels of utilization than large ones. This structure reinforces the current trend toward small, less efficient refineries and should be dropped.

This recommendation could be implemented by amendment to the regulations.

7. Eliminate the supplier/purchaser freeze upstream from the refiner level, but retain the downstream freeze: Entitlements and the buy/sell program obviate the need to freeze supplier-purchaser relationships in the provision of crude oil to refineries. The supplier/purchaser freeze has two main purposes: first, to ensure compliance with the price controls so that integrated firms cannot play games with internal transfer prices in order to circumvent the regulations; and second, to ensure an adequate allocation of crude oil to all refiners. The first problem, however, is eliminated by the crude-cost-equality device of entitlements and the second is taken care of quite satisfactorily by a revised buy/sell program as outlined above.

The freeze of supplier/purchaser relations, however, is necessary between refiners and marketers and among marketers during a shortage to ensure an adequate allocation of product to marketers who might otherwise be squeezed by integrated firms.

This recommendation would require an energy action.

8. Retain a crude oil entitlements program without a small refiner bias and establish a products entitlements program without a small refiner bias: The decision not to allow the market to allocate short supplies means that crude oil prices would be controlled during a severe supply interruption, even after 1979. Therefore, an entitlements program would continue to be necessary to equalize the cost of crude oil for all refiners. In conditions of shortage, the fact that

166

different refiners had different crude oil costs would not affect their abilities to sell output but would result in very different price or profit figures for the various companies based purely on their access to domestic crude oil supplies. The entitlements program should not be biased toward any group of refiners, however. In addition, for the reasons set forth above, the program should include nonbiased product entitlements to ensure that all savings are passed on to consumers.

Both retention of the entitlements program without a small refiner bias and establishment of a product entitlements program without a small refiner bias could be accomplished by amendment to the regulations.

Consider revising the contingency rationing plan. Section 203 (a) (1) of EPCA requires the President to produce a contingency rationing plan for approval by Congress. FEA prepared such a plan, published it for comments, and has now revised its plan. Under the legislative scheme and as proposed by FEA, rationing is viewed as the choice of last resort (to handle a severe energy supply interruption), to be used only if all other techniques for managing a shortage seem unlikely to work. FEA has interpreted the fact that rationing will be tried only after other techniques prove insufficient to mean that any rationing plan must be appended on to those other techniques, rather than serving as a substitute for them. While the task force does not take issue with the concept that rationing is a last resort approach, we do take issue with any rationing plan which fails to recognize two facts: first, like it or not, the public may demand rationing in a shortage of significant duration; second, rationing offers advantages not available with other techniques for managing a shortage. Any plan finally adopted by FEA should recognize these facts.

Fundamentally, a shortage requires that demand be lowered to equal the suddenly reduced supply, and this can be done in either of two ways. One approach is to decree how much each seller may sell and then let end users scramble to be the recipients. This is the approach taken under current price and allocation controls with respect to the general public. The other approach is to prescribe the amount end users may receive and then let sellers compete to be the suppliers. This is the opportunity offered by using rationing to manage a shortage: the benefits to consumers of competition among sellers can be preserved without undue damage to the long-term structure of the industry. To preserve these benefits, however, the rationing plan must be a *substitute* for most of the price and allocation controls, rather than an appendage to them. To promote com-

167

petition and its benefits for consumers, in other words, a rationing plan must be carefully drawn with that goal in mind.

A second basic advantage of rationing not shared by price and allocation regulations is that rationing ensures that priority uses of fuel can be served no matter how dispersed the distribution system is. The allocation system is designed to control the distribution of fuel supplies down through the wholesale level, but it cannot ensure that priorities are served among final consumer uses. To do so requires some allocation system among final users. For a rationing system to ensure that declared priorities are followed it must contain clear mechanisms for adjusting the allowed consumption of all users to the total quantity available, regardless of how small the available supplies may be.

The next section lays out the necessary ingredients of a rationing system that both ensures that priorities are served and maximizes the benefits of competition for consumers. The subsequent section critiques the FEA proposed rationing plan.

1. Outline of a contingency rationing plan which the task force recommends be considered by FEA: The first requirement for an effective rationing plan is that it be designed to limit the allowed consumption to be equal to or even slightly less than the anticipated supply of the item being rationed. Keeping the allowed consumption less than anticipated supplies allows some room for maneuver both to correct obvious hardship cases and to handle cases where actual supplies turn out to be less than anticipated. Thus, a rationing scheme must serve as a means of limiting demand.

Keeping the allowed consumption slightly below the anticipated supplies not only allows for slippage in any estimates but also permits competition to work among sellers to the benefit of consumers if the rationing plan is not merely appended to the price and allocation regulations. These latter regulations block competition within the different levels of the industry. To preserve competitive behavior and benefits, a rationing plan should be designed broadly as follows. When imposed, coupons (or ration credit accounts) would be distributed to all eligible individuals according to the priorities established in the law or spelled out in the regulations. The coupons would be good only for a specified amount of fuel and period of time for end users. The initial coupon recipients should be free to sell part or all of their allotment if they prefer. This permits a market mechanism to adjust allowed demand to differing needs among individuals while still ensuring that every eligible user can get some supplies regardless of income level.

For each gallon of fuel bought at each level of the industry, a buyer would have to turn over valid coupons to the seller. These in turn would become the necessary authorization for the seller to purchase additional supplies. There would be no time limit restrictions on a seller's redemption of coupons. The quantity of fuel that a seller could receive from a refiner or a wholesaler would be directly proportional to his proportion of the total number of coupons redeemed by final users. If a gasoline service station received its supplies from a wholesaler, the coupons would in turn serve the wholesaler as his right to purchase from refiners.

In essence, coupons would serve the same function that money does today: retailers who receive more money today can buy more supplies to sell tomorrow than can retailers who are less successful at attracting customers. Under the rationing system retailers would need both money and coupons, making coupons a special kind of money. The quantities retailers would receive would depend solely on how well final consumers liked particular stations, rather than on base period behavior.

It is important in this that the total number of coupons be slightly less than the total amount of supplies anticipated so that firms at all levels of the industry will be motivated to compete to avoid being those that have unsold supplies. If this criterion is met, such a rationing scheme would permit abolition of the freeze on supplier/purchaser relations, allocations, and all price controls beyond the refinery level. Because there might be some slippage in the goal of keeping allowed consumption below available supplies, price controls on refinery output might still be desirable.[17]

The class-of-purchaser rules would not be necessary, however. A simple requirement for a uniform price, or controls on the weighted average price, combined with the need for sellers to redeem coupons to receive supplies would be sufficient to prevent refiners from squeezing out independent marketers of gasoline. If independent marketers are the ones with coupons from final consumers, refiners will have no choice but to sell to them. And, if supplies are really larger than demand as constrained by coupons, refiners would have to balance any attempt to load price increases on independents against the possibility that the independents would be able to turn to other refiners for supplies. A refiner who made the wrong choice would find himself with the costs of storing a growing inventory of products. If ration-

_____

[17] But note that during the 1974-1975 Arab oil embargo FEA indeed reduced allowed demand to below available supply as evidenced by the growth of stocks during the period.

ing extends to all major refinery products, refinery yield programs would also not be necessary, allowing abolition of another set of regulations. The total number of coupons for each product would be known, which would provide more information to refiners than they now have about the relative mix of final demand for their products. Knowing the relative mix of final demand would thus be more than sufficient to allow them to decide how to allocate their output.

2. Shortcomings of the FEA contingency rationing plan: The FEA contingency rationing plan as originally proposed and as modified according to the plan's management summary does not resemble the above plan in certain important respects. Its most notable difference is that it is explicitly intended to be appended to the price and allocation regulations, thus merely ending the scramble among final consumers to see who gets how much of what limited supplies are available. Thus, when activated, FEA's rationing plan would simply add another layer of regulation while maintaining a whole set of unnecessary and costly regulations. More importantly, adoption of the FEA plan would foreclose the opportunity even in a shortage to try to preserve the benefits of competition for consumers.

Several features of the FEA plan are responsible for its inability to serve such a purpose. In addition to the stated intentions of being an addition to, not a substitute for, the price and allocation regulations, the plan itself would need modification to serve as a substitute. The first would be an explicit recognition of its role as a tool to bring allowed demand into balance without available supplies. The preamble to the present FEA plan reflects a reluctance to use rationing as a means of reducing total demand to the level of available supply, stating: "The proposed rationing plan is not a system to reduce national demand." This reluctance is further revealed in the section establishing ration credits for business and government agencies. The plan proposes to establish these accounts allowing firms fixed proportions of base year usage regardless of the total available supply, with the adjustment to total supply coming in the distribution of ration coupons to individuals. There is no mechanism for adjusting ration credit accounts if the total available supply is less than the fixed allotments provided for in the rationing plan.

The task force recommends that FEA consider modifications to convert its plan into a substitute for price and allocation controls. Even if regarded as a technique of last resort, the contingency rationing plan adopted should take full advantage of every opportunity to achieve statutory goals at minimum economic cost. If the benefits to consumers from competition can be preserved while ensuring that windfall gains are not collected by producers, this surely should be done.

The recommended modifications to the contingency rationing plan could be implemented by amendment of the program.

Recommended Changes in FEA Procedures

Recommended Changes in the Compliance Program. The following recommendations for modification of the compliance program contemplate the enforcement, in a shortage situation, of simple, unambiguous regulations. In such an emergency situation, given the clarity of the rules, vindication of the public interest in its safety, health, and welfare would justify a vigorous enforcement effort.

The task force has noted at least one occasion in which FEA has proposed granting to a significant class of firms (those refiners which erroneously applied the cost passthrough order of recoupment rules) retroactive exception relief, potentially waiving $1.3 billion in overcharges.

Retroactive relief of this magnitude should not be used by FEA in the future as a means of nullifying regulatory violations. The practice fosters two fundamental problems: first, it undermines any possibility of effective enforcement based upon general deterrence; second, it creates inequities by penalizing firms which make a good faith effort to comply and by rewarding firms which do not.

In addition, given the recommended changes to the regulatory framework, the agency ought to have an enforcement program which will effectively deter as well as remedy violations of the regulations. The task force recognizes that the imposition of criminal penalties may be appropriate only in the case of very serious or flagrant violations. Consequently, in the majority of cases in which civil remedies are appropriate, the remedy sought must be tailored not only to alleviate the effects of the violation but also to deter further disregard of the program. This approach would require, in many cases, the imposition of substantial civil penalties. Likewise, if the task force recommendation that banks be eliminated is not adopted, FEA should confine the use of bank adjustments as a remedy to violations of the bank regulations in which no improper costs have been charged purchasers.

In sum, only if FEA significantly increases the deterrent effect of its compliance program will the full public benefit of the task force substantive recommendations be realized.

Recommended Modifications of the Exception Process. The exception process has evolved beyond its original function of providing "exceptional" relief from unexpected or unintended effects of FEA regulations. Today, the process additionally provides an alternate mechanism by which existing FEA regulations may be revised through a process of case-by-case development rather than through the formal rulemaking processes. Yet, FEA appears not to have recognized formally this emerging exception function. As a result, the exception process presently is unable to utilize fully the flexibility inherent in case-by-case rulemaking, while imposing burdens on many who seek relief from FEA regulations, creating vested interests in the regulatory program, and modifying regulations without the ability to consider alternative regulatory solutions.

The task force views the continuation of the present situation as unsatisfactory. It therefore recommends that FEA make a formal decision about whether the exception process should be granted full authority to make revisions in regulations that operate in unintended ways or become outmoded, or whether exception decisions should be limited to providing genuinely exceptional relief in unique circumstances and to alleviating serious hardship in time of emergency.

The task force believes that the preferred solution is to place more reliance on the formal regulatory development process, particularly in an agency with a temporary regulatory mandate. Expansion of the exception process to encompass plenary case-by-case rulemaking would bring FEA practice into a pattern adopted by many established regulatory agencies, a development which would be inconsistent with the nature of the FEA regulatory scheme. The decision, however, is one which must be squarely faced by the agency.

If FEA decides to utilize the exception process as part of the regulatory development mechanism, its scope must be expanded. Utilization of the exception process to establish rules of general applicability would require several changes in existing exception practices. The task force has not made a comprehensive study of all the required procedural changes, nor has it attempted to recommend organizational changes in FEA which might have to be made to accommodate an expanded role for the exception process in policy making. Likewise, the task force notes that the ability of FEA to establish rules—as opposed to "exceptions" to rules—by adjudicatory techniques is not clearly spelled out by the Federal Energy Administration Act. Thus, further legal research and possible statutory amendment may be necessary before this proposal is adopted.

If FEA decides to expand the role of the exception process so as to permit it to make rules of general application, the revised process should at a minimum have the following attributes:

1. Decisions should at least be able to state a specific modification, amendment, or exception to a given rule; the regulations should make specific cross-reference to the decision.
2. Relief should run to all similarly situated parties without the need for each to file an exception or join as a class. There need be no time limit on relief; poorly drafted regulations should be changed permanently to "correct" any problems.
3. Where relief requires the evaluation of individual circumstances, the decision announcing the adoption of a change should state that other parties wishing to obtain similar relief need only submit specified data and need not reargue the need for the general rule.
4. When FEA determines that a particular application will likely involve the determination of a rule of general application, the agency should solicit comments from all interested parties and state the possible rules which it may consider in granting relief.

If FEA decides to utilize the exception process more narrowly, relief granted should be limited in scope, and the exception function should be linked more closely to the formal rulemaking process. The task force strongly favors limiting the exception process to providing case-by-case exceptions to remedy individual problems, rather than establishing rules of general application. Such a restricted role is appropriate to the temporary nature of the program. If FEA takes this approach and decides to limit the exception process, the task force recommends the following changes in FEA procedures:

1. "Serious hardship" relief should be granted only during a severe supply interruption for the purpose of preventing the failure of firms as the result of arbitrary rules such as fixed pricing and allocation base periods and supplier/purchaser freezes. In this regard, it should be noted that the task force has recommended the elimination of such arbitrary rules during nonemergency periods.
2. "Serious hardship" and "gross inequity" should be eliminated as separate grounds for exception relief during nonemergency periods. There should be a single standard pursuant to which exception relief will be granted when, due to unusual or unforeseen factual circumstances, particular FEA regulations have caused unintended hardship to a single firm or very limited group of firms. Exception relief should not be granted to encourage additional investment in particular projects absent a showing of actual harm to the applicant.
3. The relief given under the revised "gross inequity" standard should be sufficient, but no more than sufficient, to remedy the

unforeseen or unintended harm. In particular, care should be taken to avoid establishing a type of relief which operates to maintain traditional profit levels at a time when no severe supply shortage exists.

4. In those cases in which the Office of Exceptions and Appeals or the Office of Regulatory Programs determines that relief should be granted and/or measured by reference to a formula or other routine standard, FEA should promulgate such standards through the rulemaking process. The agency should establish a mechanism distinct from an exception application for the routine granting of relief provided by such regulations.

5. When it appears from an exception application that an FEA regulation has inadvertently failed to take into consideration the ongoing operation of a particular type of business activity, exception relief should be granted on a continuing basis to permit the applicant to behave in a manner analogous to that of the firms for which the regulation was specifically designed. If it is likely that more than a very few firms are affected by the defective regulation, a rulemaking inquiry should be undertaken to determine the appropriate regulatory remedy.

6. Major policy decisions, such as those to assist a particular class of firms in maintaining their historic rates of profitability, should be made only through the rulemaking process. Consequently, when the Office of Exceptions and Appeals receives a meritorious request for an exception on grounds which are of general application or which involve a request for relief not appropriate to the exceptions process, the matter should be referred to the Office of Regulatory Programs for a rulemaking. Procedures should be established, however, to stay the effect of the regulation as to the applicant or to grant a temporary exception, pending resolution of the rulemaking.

Recommended Changes in Regulatory Development Procedures.* The recommendations that follow are directed specifically at the process of developing allocation and price regulations, the major focus of task force investigations. However, we believe the proposed development concept may be applicable to other areas of the agency's responsibility, including conservation, energy resource development, and strategic storage.

---

* Editor's note: Only the task force's summary of its recommendations for changes in regulatory development procedures has been included.

It is the opinion of the task force that the establishment of a standard procedure for regulatory development and a clear definition of primary functional responsibility among offices is essential for correction of the problems outlined above. Therefore, our principal recommendation is that the agency create a regulatory development executive committee comprised of the deputy administrator, the general counsel, and appropriate assistant administrators; and the appointment of a regulatory development coordinator. The committee would meet regularly, under the direction of the deputy administrator, make all significant regulatory policy decisions (subject to the administrator's concurrence), assign responsibilities to offices, and set priorities and deadlines. The coordinator would prepare agendas for committee meetings, record minutes of meetings, and monitor schedules, deadlines, and the circulation of issue papers and drafts between offices. The coordinator would not assume any policy-making responsibilities. The recommended procedure allows for any office to initiate a regulatory proposal, assures prompt and well-informed decisions by the committee on whether to proceed with proposals, provides for appropriate assignment of regulation-drafting responsibilities, and furnishes opportunities for all concerned offices to review and comment. By using this procedure, the office with primary program responsibility would be responsible for drafting applicable regulations and for securing necessary analysis of economic and enforcement issues and would be held accountable for the end product.

Evaluation criteria should, at a minimum, include consideration of statutory objectives, manner of enforcement, and economic impact. In evaluating the statutory objectives, the agency should be certain not only of the legal sufficiency of the proposed regulation (a determination well-suited to the Office of General Counsel) but also of its ability to promote the policies and programs mandated by the statutes; for example, protect competition, provide for economic efficiency, and minimize unnecessary interference with market mechanisms. To determine the enforceability of a proposed regulation, the agency should examine whether it is understandable, whether it is adaptable to existing business practices, and whether it permits compliance monitoring. In assessing the economic impact of each proposed regulation and the viable alternatives, the analyses should be performed early in the process and should cover such items as effect on consumers, competition, efficiency, and domestic production. The analyses should include any required inflation impact statements.

Finally, the task force recommends that offices which play significant roles in the development of regulations be centrally located and that recruitment and training be aimed at improving employee knowledge of the regulations and the industry.

For any improvements to the regulatory development process to be successful, the full endorsement and continuing support of the administrator is essential.

GLOSSARY

Agency	Reference to the Federal Energy Administration.
Aggrieved party	A person whose statutorily protected interest is adversely affected by an order or interpretation issued by FEA.
Agricultural user	Refers to activities which are enumerated in section 211.51.
Allocation fraction	A fraction calculated as described in section 211 (b), which each supplier uses to apportion his allocable supply among purchasers.
Allocation level	The proportion of an end user's base period volume, adjusted base period volume, or current requirement, as appropriate, that his supplier is authorized to deliver to him if sufficient amounts are available. The level varies with the class of purchaser or end use to be made of the allocated substance.
Arab oil embargo	The oil embargo imposed by the Arab states on nations friendly to Israel during the period October 1973 to April 1974.
Aviation fuels	Refers to kerosene-type or naphtha-type aviation fuels or to aviation gasoline.
Banked costs	An accumulation of allowable product and non-product cost increases which were not passed through to purchasers in the form of higher prices.
Barrel of crude/product	Equivalent to forty-two gallons.

177

Base period	The historical period designated within FEA's regulations dependent upon the product involved (crude oil, residual fuel oil, or refined petroleum products).
Base price	Until its deletion April 12, 1976, it was determined by section 212.111.
Base period supplier	See Supplier.
Branded independent marketer	A firm which is engaged in the marketing or distributing of refined petroleum products pursuant to:

(a) An agreement or contract with a refiner (or a firm which controls, is controlled by, or is under common control with such refiner) to use a trademark, trade name, service mark, or other identifying symbol or name owned by such refiner (or any such firm), or

(b) An agreement or contract under which any such firm engaged in the marketing or distributing of refined petroleum products is granted authority to occupy premises owned, leased, or in any way controlled by a refiner (or firm which controls, is controlled by, or is under common control with such refiner), but which is not affiliated with, controlled by, or under common control with any refiner (other than by means of a supply contract, or an agreement or contract described in paragraph (a) or (b) of this definition), and which does not control such refiner.

Bulk terminal	A facility which is primarily used for the marketing of gasoline, kerosene, and distillate and residual fuel oils and which (a) has total bulk storage capacity of 2,100,000 gallons or (b) receives its petroleum products by tanker, barge, or pipeline.
Bulk user (purchaser)	Any firm which is an ultimate consumer which, as part of its normal business practices, purchases or obtains motor gasoline from a supplier and either (a) receives delivery of that product into a storage tank substantially under the control of that firm at a fixed location, (b) with respect to use in agricultural product, receives delivery into a storage tank with a capacity not less than 50 gallons sub-

	stantially under the control of that firm, or (c) receives delivery of that product for use in cargo, freight, and mail hauling by truck.
Butane	A hydrocarbon whose chemical composition is predominately C_4H_{10}, whether recovered from natural gas or crude oil.
Buy/sell program	FEA program which provides for the mandatory allocation of crude oil produced in or imported into the United States.
Capital	A term denoting any physical goods which are produced, not because of any intrinsic value to consumers, but because they are necessary, directly or indirectly, to produce other goods and services that satisfy consumer wants.
Catalytic converter	A pollution-control device on a motor vehicle.
Changed circumstances	Includes such factors as plant expansion, changed traffic patterns, closed retail sales outlets which have increased demand on remaining outlets, changes in the local economy, unusual seasonal fluctuations, new population, and industrial growth.
Class of purchaser	Purchasers to whom a person has charged a comparable price for comparable property or service pursuant to customary price differentials between those purchasers and other purchasers.
Crude oil	A mixture of hydrocarbons that existed in liquid phase in underground reservoirs and remains liquid at atmospheric pressure after passing through surface separating facilities. "Crude oil" includes condensate recovered in associated or nonassociated production by mechanical separators, whether located on the lease, at central field facilities, or at the inlet side of a gas processing plant.
Crude oil entitlements	See Entitlements.
Crude oil runs to stills	The sum of the total number of barrels of crude oil input to distillation units processed by a refiner and measured in accordance with Bureau of Mines

179

Form 6-1300-M. The volume of a refiner's crude oil runs to stills also includes inputs to distillation units of plant condensate produced in and imported from Canada and synthetic crude oil made from tar sands and imported from Canada.

Current requirements

The supply of an allocated product needed by an end user or wholesale purchaser-consumer to meet its present supply requirements for a particular use of that product but does not include any amounts which the end user or wholesale purchaser-consumer (a) purchases or obtains for resale, (b) accumulates as an inventory in excess of that purchaser's customary inventory maintained in the conduct of its normal business practices, or (c) uses in excess of the supply necessary to meet present supply requirements as constrained by the implementation of the energy conservation program required in section 211.21.

Dealer

See Retailer.

Desulfurization

Removal of sulfur or sulfur compounds from crude oil or its products.

Diesel fuel

The petroleum fraction used as a fuel in diesel or compression ignition engines. Various qualities are marketed depending on the type of engine operated. The most important characteristic of diesel fuel, particularly 1-D and 2-D, is its ignition quality, since this controls its performance in the engine. Ignition quality is determined in an engine as the "cetane number." Volatility also affects engine performance and is generally controlled by the distillation range. Most diesel fuels fall in the range of thirty to sixty-five in cetane numbers.

Distillate fuel

A product of distillation, or the liquid condensed from the vapor driven off in the still. Gasoline, naphtha, solvents, kerosene, number 1 and number 2 heating oil, diesel fuels, jet fuels, propane, butane, and lubricating oils are all examples of distillation since they are the result of distillation of crude oil. Middle distillates are the intermediate ranges of fuel such as kerosene, home heating oils, range and stove oil, and diesel fuel.

Distributor   A person, other than a manufacturer or retailer, to whom a consumer product is delivered or sold for purposes of distribution in commerce.

Downstream   Starting with one stage of a sequential production process and encompassing all subsequent stages.

Economies of scale   Decreases in the cost of production from any type of plant which are associated with increases in the size of the plant.

Embargo   See Arab oil embargo.

End user   Any firm which is an ultimate consumer of an allocated product other than a wholesale purchaser-consumer.

Entitlements   For a particular month, the right of the refiner owning the entitlement to include one barrel of deemed "old" oil in its adjusted crude oil receipts that month.

Exception   The waiver or modification of the requirements of a regulation, ruling, or generally applicable requirement under a specific set of facts.

Exemption   Release from the obligation to comply with any part or parts, or any subpart of the regulations.

Feedstock   Use of crude oil, residual fuel oil, and refined petroleum products for processing in a petrochemical plant.

Firm   Any association, company, corporation, estate, individual, joint venture, partnership, sole proprietorship, or any other entity however organized including charitable, educational, or other eleemosynary institutions, and the federal government including corporations, departments, federal agencies, and other instrumentalities, and state and local governments. The FEA may, in regulations and forms issued in this part, treat as a firm: (a) a parent and the consolidated and unconsolidated entities (if any) which it directly or indirectly controls, (b) a parent and its consolidated entities, (c) an unconsolidated entity, or (d) any part of a firm.

Gas and go	Gas station which has no repair services, sells gasoline only.
Gasoline	See Motor gasoline.
Gathering systems	Series of small diameter pipes connecting all producing leases in an area to a common point so that crude oil can be transported to a trunk pipeline.
General refinery products	All covered products other than number 2 oils, aviation jet fuel, gasoline, and crude oil.
Gravity/price differential	A price distinction; based on a difference in densities of crude oil (the higher the API specific gravity measurement, the more valuable the crude).
Gross margin	Amount of revenues a firm realizes over the cost of a particular product for the sale of the particular product.
"H" factor	An element of the refiner price rule formula which represents the reallocation of increased measured product costs.
Heating oil, home	See number 2 heating oil.
Import	Entry into the United States for consumption.
Import fee	The charge imposed on crude petroleum and petroleum products which enter into the United States for consumption.
Imputed price	The price that would be paid domestically for imports of petroleum products in the presence of product entitlements.
Independent marketer	Either a branded independent marketer or a nonbranded independent marketer.
Independent refiner	A refiner which (a) obtained, directly or indirectly, in the calendar quarter which ended immediately prior to November 27, 1973, more than 70 percent of its refinery input of domestic crude oil or 70 percent of its refinery input of domestic and imported

crude oil from producers which do not control, are not controlled by, and are not under common control with such refiner; and (b) marketed or distributed in such quarter and continues to market or distribute a substantial volume of gasoline refined by it through branded independent marketers or nonbranded independent marketers.

Input
Anything that is used in the production of something else.

Jobber
A firm that buys and stores petroleum products at or near consumption centers and sells and delivers petroleum products to resellers, retailers, and end users.

Kerosene
Any jet fuel, diesel fuel, fuel oil, or other petroleum oils derived by refining or processing crude oil or unfinished oils, in whatever type of plant such refining or processing may occur, which has a boiling range at atmospheric pressure which falls completely or in part between 400° and 500° F.

Lessee dealer
An independent seller, generally of gasoline, who rents a service station from his supplier.

License fee
An import fee. See Import fee.

Majors
The ten or fifteen largest oil companies. They are often recognized to be the integrated oil companies that produce, refine, transport, and market under their own brand in twenty or more states.

Marketer
Any firm which purchases, receives through transfer, or otherwise obtains (as by consignment) an allocated product and resells or otherwise transfers it to other purchasers without substantially changing its form.

Middle distillates
Any derivatives of petroleum including kerosene, home heating oil, range oil, stove oil, and diesel fuel, which have a 50 percent boiling point in the ASTM D86 standard distillation test falling between 371° and 700°F. Products specifically excluded from this definition are kerosene-base and naphtha-base jet fuel; heavy fuel oils as defined in

VV-F-815C or ASTM D-396; grades number 4, 5, and 6; intermediate fuel oils (which are blends containing number 6 oil); and all specialty items such as solvents, lubricants, waxes, and process oil.

Motor gasoline	A mixture of volatile hydrocarbons, suitable for operation of an internal combustion engine, whose major components are hydrocarbons with boiling points ranging from 140° to 390°F and whose source is distillation of petroleum and cracking, polymerization, and other chemical reactions by which the naturally occurring petroleum hydrocarbons are converted to those that have superior fuel properties.
Naphthas	All petroleum fractions, not otherwise defined as aviation fuels, gasoline, or special naphthas, made up predominantly of hydrocarbons whose boiling point falls within the temperature range of 85° to 430°F.
Natural gas	A natural hydrocarbon gas composed of a variety of gases including methane, othane, butane, and propane. It comes from the ground with or without accompanying crude oil. It is generally much higher in heat content than manufactured gas. It is used as the raw material in the petrochemical industry in the manufacture of fertilizer and cellophane.
Natural gasoline	Those liquid hydrocarbon mixtures containing substantial quantities of pentanes and heavier hydrocarbons, which have been extracted from natural gas. It is used as a blending stock since it is volatile and aids in engine starting. It has good antiknock qualities and contains butane, propane, and other fractions.
Net margin	Amount of revenues a firm realizes over the cost of a particular product for the sale of the particular product, adjusted for additional costs incurred.
New oil	The domestic crude oil produced from a property in a specific month less the amount produced from the same property in the same month/1975.

Nonbranded independent marketer	A firm which is engaged in the marketing or distribution of refined petroleum products, but which (a) is not a refiner; (b) is not a firm which controls, is controlled by, is under common control with, or is affiliated with a refiner (other than by means of a supply contract); and (c) is not a branded independent marketer.
Nonproduct cost increases	Increases in operating costs of refiners, retailers, and resellers. Nonproduct costs for refiners are listed at 10 C.F.R., section 212.83(c)(1)(i)(E).
Number 2 heating oil	Heating oil grade number 2 as defined in American Society for Testing and Materials (ASTM). This grade of oil is used in most home burners that have central heating and in many medium-capacity commercial industrial burners where its ease of handling sometimes justifies its higher cost over the residual fuels.
Octane	Measure of combustion qualities of a grade or blend of motor gasoline.
Octane (clear)	Rating of a gasoline obtained by a process which does not include lead or other molecular components.
Old oil	(a) Prior to February 1, 1976, the total number of barrels of crude oil produced and sold from a property in a specific month, less the total number of barrels of new crude oil for that property in that month, and less the total number of barrels of released crude oil for that property in that month; (b) effective February 1, 1976, the total number of barrels of crude oil produced and sold from a property in a specific month, less the total number of barrels of new crude oil for that property in that month.
Order of recoupment	Order in which a refiner recovers increased product and nonproduct costs.
Other products	Allocated products including benzene, toluene, mixed xylenes, hexane, lubricants, greases, special naphthas (solvents), lubricant base stock oils, and process oils.

Out-of-pocket cost	The cost of any input, the use of which increases as output increases, for any given plant.
Output	The volume of a good that is produced.
Peak-shaving	The use of propane or butane mixture to supplement supplies of pipeline gas for distribution by gas utilities during periods of high demand.
Petrochemicals	Chemicals made from components of crude oil and/or natural gas and listed in section 211.12(k).
Petroleum or petroleum products	Crude oil, residual fuel, or any refined petroleum product (including any natural gas liquid or any natural gas liquid product); includes all petroleum fuels, lubricants, and specialties.
Phase IV	Period of the Economic Stabilization Program which began in August 1973 and for petroleum products ended when FEO came into existence (December 4, 1973).
Price differential	A price distinction; based on a discount, allowance, add-on premium, and an extra based on a difference in volume, grade, quality, or location or type of purchaser, or a term or condition of sale or delivery.
Price elasticity	An economic term referring to the percentage change in demand for a product for each percentage change in price.
Price refund	Direct return of price overcharges to those who were overcharged.
Price rollback	Reduction in price charged which is below the maximum lawful selling price to reflect a return of overcharges.
Producer	A firm which produces propane in a refinery, natural gas processing plant, or fractionating plant, or imports more than 2 million gallons per year, including firms which own natural gas and have their gas processed for their account by others but retain title.
Product cost	Direct cost of crude oil or purchased product.
Product fees	See Import fees.

Propane	The chemical C_3H_8 in its commercial form including propane-butane mixes and other mixtures in which propane constitutes greater than 10 percent of the mixture by weight. Excluded from the definition are mixtures containing propane (other than propane-butane mixes) when such mixtures are used for refinery fuel use.
Property	The right to produce domestic crude oil, which arises from a lease or from a fee interest. A producer may treat as a separate property each separate and distinct producing reservoir subject to the same right to produce crude oil, provided that such reservoir is recognized by the appropriate governmental regulatory authority as a producing formation that is separate and distinct from, and not in communication with, any other producing formation.
Prorationing	Process of apportioning production among a number of wells such that the total production does not exceed some specified amount.
Pump prices	Prices charged by retail gasoline outlets.
Refined petroleum products	Gasolines, kerosene, middle distillate (including number 2 fuel oil), LPG, refined lubricating oils, or diesel fuel.
Refineries	Those industrial plants, regardless of capacity, processing crude oil feedstock and manufacturing refined petroleum products, except when such plant is a petrochemical plant.
Refiners	Those firms that own, operate, or control the operation of one or more refineries.
Regions	One of the ten regions served by FEA's regional offices.
Regional office	A regional office of FEA. The regional offices are located in Boston, Massachusetts; New York, New York; Philadelphia, Pennsylvania; Atlanta, Georgia; Chicago, Illinois; Dallas, Texas; Kansas City, Missouri; Denver, Colorado; San Francisco, California; and Seattle, Washington.

Regulations	Current and past regulations in effect for FEA.
Released oil	An amount of crude oil produced from a property in a particular month prior to February 1, 1976, which is equal to the total number of barrels of new crude oil produced and sold from that property in that month. The amount of released crude oil for a property in a particular month shall not exceed the base production control level for that property in that month, and shall not include any number of barrels not certified as such pursuant to the provisions of section 212.131(a)(1) within the consecutive two-month period immediately succeeding the month in which the crude oil is produced and sold, except where such recertification is explicitly required or permitted by FEA order, interpretation, or ruling.
Rents	Returns to the owner of a commodity that is fixed in quantity and scarce, the value of which is determined by the commodity's capacity.
Rent controls	Limitations by the state on rents.
Reseller	Any person, firm, corporation, or subdivision thereof that carries on the trade or business of purchasing any allocated substance and reselling it without substantially changing its form.
Residual fuel oil	The fuel oil commonly known as: (a) number 4, number 5, and number 6 fuel oils; (b) Bunker C; (c) Navy Special Fuel Oil; and all other fuel oils which have a 50 percent boiling point over 700°F in the ASTM D-86 standard distillation test.
Retail sales outlets	A site on which a supplier maintains an ongoing business of selling any allocated product to end users or wholesale purchaser-consumers.
Retailer	Any person, firm, corporation, or subdivision thereof that sells any allocated substance directly to any end user.
Rulemakings	See Regulations.
Sanctions	Penalties.

Secondary recovery techniques	Method of recovering additional energy from a reservoir by injecting fluid into the reservoir.
Small refiner	(a) Generally, a refiner whose total refinery capacity (including the refinery capacity of any firm which controls, is controlled by, or is under common control with such refiner) does not exceed 175,000 barrels per day; (b) for RARP, a refiner whose total refinery capacity does not exceed 75,000 barrels per day.
Small refiner bias	Any tendency caused by FEA regulations which grants to small refiners benefits greater than such refiners would reap in a free market.
Sour crude	Crude oil containing sufficiently large quantities of sulfur and sulfur compounds so as to require chemical treatment for removal.
Special products	Gasoline, number 2 heating oil, and number 2-D diesel fuel.
Spot market price	A market in which prices are quoted for a commodity on the basis of immediate sale and delivery.
State set-aside	With respect to a particular prime supplier, the amount of an allocated product which is made available from the total supply of a prime supplier pursuant to section 211.17 for utilization by a state to resolve emergencies and hardships due to fuel shortages. The state set-aside amount for a particular month and state is calculated by multiplying the state set-aside percentage level by the prime supplier's estimated portion of its total supply for that month which will be sold into that state's distribution system for consumption within the state. The initial state set-aside percentage level for an allocated product is specified in the appropriate subpart for that product, but is subject to change by notice of the FEA.
Stripper wells	A well producing on the average of ten barrels or less of crude oil per day.

Supplier	Any firm or any part or subsidiary of any firm other than the Department of Defense which presently, during the base period, or during any period between the base period and the present supplies, sells, transfers, or otherwise furnishes (as by consignment) any allocated product or crude oil to wholesale purchasers or end users, including, but not limited to, refiners, natural gas processing plants or fractionating plants, importers, resellers, jobbers, and retailers.
Supplier/ purchaser relationship	Relationship between a supplier and his wholesale purchasers and/or end users during the designated base period of each allocated product.
Sweet crude	Crude oil containing so little sulfur as to render unnecessary any chemical treatment for the removal of sulfur or sulfur compounds.
Tertiary recovery techniques	Methods of recovering additional energy from a reservoir by injecting chemicals into the reservoir.
Throughput	Volume of crude oil, unfinished oil, and natural gas liquids refined during the time period specified.
Unit cost	See Weighted average unit cost.
Utilization	Allows an oil field to be produced as a single property or "unit" without regard to drainage.
Unusual growth adjustment	An adjustment to base period volume based upon a comparison of the base period volume purchased from all suppliers in 1972 and the volume purchased from all suppliers in 1973. Growth which exceeds 10 percent for motor gasoline or 5 percent for any other allocated product is defined as "unusual growth."
Upstream	Starting with one stage of a sequential production process and encompassing all prior stages.
Utilization level	The ratio of a plant's current output to the maximum output which the plant is designated to produce.

Vacuum distillation	A distillation process accomplished in a vacuum in order to reduce the process temperature sufficiently to avoid decomposition of the oil.
Weighted average unit cost	Total cost of product purchased divided by the quantity purchased.
Wholesale purchaser	A wholesale purchaser-reseller or wholesale purchaser-consumer, or both.
Wholesale purchaser-consumer	Any firm that is an ultimate consumer which, as part of its normal business practices, purchases or obtains an allocated product from a supplier and receives delivery of that product into storage substantially under the control of that firm at a fixed location and purchased or obtained more than 20,000 gallons of lubricants, 10,000 pounds of greases, or 55,000 gallons of any other product in any completed calendar year subsequent to 1971.
Wholesale purchaser-reseller	Any firm which purchases, receives through transfer, or otherwise obtains (as by consignment) an allocated product and resells or otherwise transfers it to other purchasers without substantially changing its form.
Windfall profits	Any profits which could not have been foreseen or planned for.

Note: The above definitions of terms are based primarily upon definitions found within the FEA Guidelines.

ABBREVIATIONS

ASTM	American Society for Testing and Materials
BPCL	Base Period Control Level
CLC	Cost of Living Council
DOD	Department of Defense
ECPA	Energy Conservation and Production Act
EIA	Energy Information and Analysis
EPA	Environmental Protection Agency
EPAA	Emergency Petroleum Allocation Act of 1973 or 1975
EPCA	Energy Policy and Conservation Act
EPO	Energy Policy Office
ESECA	Energy Supply and Environmental Coordination Act
ESP	Economic Stabilization Program
FEA	Federal Energy Administration
FEAA	Federal Energy Administration Act of 1974
FEO	Federal Energy Office
FPC	Federal Power Commission
FY	Fiscal Year
GAO	General Accounting Office
ICC	Interstate Commerce Commission
IIS	Inflation Impact Statement
IRS	Internal Revenue Service
MER	Maximum Efficiency Rate
MOIP	Mandatory Oil Import Program

NDCOSR	National Domestic Crude Oil Supply Ratio
NGL	Natural Gas Liquid
OAPEC	Organization of Arab Petroleum Exporting Countries
OEI	Office of Economic Impact
OGC	Office of General Counsel
OIP	Oil Import Program
OMB	Office of Management and Budget
OPEC	Organization of Petroleum Exporting Countries
ORD	Office of Regulation Development
ORM	Office of Regulation Management
ORP	Office of Regulatory Programs
OSHA	Occupational Safety and Health Administration
P&A	Office of Policy and Analysis
PAD	Petroleum Administration for Defense
PPE	Office of Policy and Program Evaluation
RARP	Refinery Audit and Review Program
SEC	Securities and Exchange Commission
SIC	Standard Industrial Classification
SNG	Synthetic Natural Gas
UPS	United Parcel Service